D0853476

The Seas in Motion

The Seas in Motion

F. G. Walton Smith

AN INTERNATIONAL OCEANOGRAPHIC FOUNDATION SELECTION

Thomas Y. Crowell Company
New York · Established 1834

 Published simultaneously in Canada by Fitzhenry & Whiteside Limited, Toronto.

Designed by Visuality

Manufactured in the United States of America

ISBN 0-690-72329-6

1 2 3 4 5 6 7 8 9 10

Library of Congress Cataloging in Publication Data

Smith, Frederick George Walton
The seas in motion.

"An International Oceanographic Foundation selection."
1. Ocean waves. 2. Tides. 3. Ocean currents. I. Title.
GC201.S65 551.4′7 72–83772
ISBN 0–690–72329–6

Foreword

The dynamic movements of the ocean are of interest and importance to all who use the ocean or its borders for recreation or livelihood. Yet the popular books and articles on waves and tides leave much to be desired, through a lack of accuracy, a superficial approach, or a failure to explain the underlying causes for many of the curious phenomena involved. This book is an attempt to strip the subject of its sometimes complex mathematics and make the movements of the sea understandable without sacrificing accuracy and comprehensiveness. First of all, the sailor, professional or amateur, who has a firsthand empirical knowledge of ocean movements and thereby a great respect for their potential dangers, has also a strong curiosity about their fundamental causes. The more he understands these causes, the better he may use his own valuable experience. The angler as much as the commercial fisherman has the same curiosity and may welcome an understanding that goes beyond experience alone. To these may be added the skin divers, whose numbers have increased so tremendously in recent years. The average tourist or seaside vacationer or the amateur shell collector has less urgent need for an exact knowledge of the seas in motion, but he is still usually very curious, as evidenced by the many letters of inquiry that have come to the author's desk during the past thirty years.

It is hoped that this book may also be a useful introduction to the oceanographic student whose training in higher mathematics is as yet inadequate for the easy reading of advanced texts in physical oceanography. The professionals, of course, the marine biologist, the marine chemist and physicist, the meteorologist, and the geologist, are all concerned with the movements of the sea, since waves, tides, and currents are very much involved in the problems they seek to solve. The engineer, concerned with the safe design of marine structures or the design of surface or subsurface ships, has his own special interests.

There has been no attempt to do more than indicate a few of the applications of a knowledge of waves, tides, and currents to the varied interests of the reader. It is hoped, however, that this will serve as a background for a better understanding of the subject and a stimulus to further reading.

Acknowledgments are gratefully accorded to those colleagues who have given advice and assistance in the task of providing a physical meaning to mathematical theories. In particular the author is grateful to Dr. Eric Kraus, Dr. Claes Rooth, and Dr. Walter Duing.

Contents

1. *Introduction*

The surface of the ocean is never at rest. To a casual observer, its motions are sometimes violent and sometimes calm, but always variable and hard to predict. Thus any discussion of the seas in motion must attempt to unravel the confused welter of storm seas and reduce them to an understandable and predictable pattern, as well as to explain the reasons for dangerous, freak waves and waves that have been reliably recorded as having a height of more than 100 feet. The discussion must give some idea of the source of power that enables a wave to move a mass of masonry weighing thousands of tons. It must also attempt to explain why tides vary from day to day; why there are virtually no tides in some places, whereas in others they range vertically more than 50 feet from low to high water; why tidal currents may be absent altogether at one locality but run at well over 10 knots at another; and why in some places the tide rises daily and, in others, twice daily. There are also questions to be answered about the vagaries of the Gulf Stream, whose burden of more than 85,000,000 tons of water per second travels on a course that constantly varies. These are very practical questions for many people. Fortunately, in most cases a practical answer can be given without becoming too involved in mathematical complexities.

To consider sea changes, we must start with a basic understanding of the nature of water itself. It is a familiar fact that the oceans cover the greater part of our planet. It is less well known that only 8 percent of the water of this watery planet is in the air or in lakes and rivers or in the form of ice. The other 92 percent is in the oceans. Least well known is that water is a unique fluid. Because of its amazing properties it is the only fluid able to support life on earth. But it also has properties that enable it to store energy—properties that indirectly account for the great storm waves, the ocean currents, and the great upwellings of water from the deep that create the rich fishing areas of the world.

The restless surface that separates the sea from the atmosphere is where the great exchanges of energy take place. There the sun's radiant energy enters the sea, where most of it is stored in the form of water heat and generates thermal currents, both horizontal and vertical. Some of the energy, however, in the form of water vapor, is passed back into the atmosphere, where it is temporarily stored. The energy is eventually released in considerable amounts when the water once more condenses as rain. In this way it provides the piston power of the great wind engine. It is no accident that hurricanes and typhoons, the greatest storms on this ocean planet, occur at sea. Over the land, deprived of water vapor, their vital fuel, they falter and die.

The exchange of energy across the surface boundary does not cease with the passage of energy from sea to air. In a constant exchange back and forth, energy from the atmosphere is transferred to the sea when winds generate waves of all sizes at the surface. The energy of great waves is readily apparent to those who travel by sea in ships, and most of all to those who race in small sailboats across whole oceans. But it is also apparent to those who ride the great surf waves of northern Oahu or watch storm crests break below them as they stand upon the cliffs.

The wind also plays a major part in generating the currents that girdle the oceans, carrying heat from the tropics to the poles and so modifying the climate of neighboring lands. These currents may aid or obstruct navigation, and their movements may alter the characteristics of the waves that ride upon them. And there are other currents running deep below the surface and still others carrying water vertically between the surface and the depths—all with profound effects upon sea life. In general, then, the major motive power of currents and waves is the wind. Thus, since energy from the sea helps to set the atmosphere in motion and the winds of the atmosphere, in turn, provide much of the power that sets the sea in motion, the energy of waves and currents is to a considerable extent derived originally from the sea itself.

This relationship is a major reason for the present increasing scientific interest in air-sea interaction.

The sun's energy is not the only energy continuously working on the ocean. In addition, the energy derived from the movements of the earth, sun, and moon generates tidal fluctuations. These fluctuations have the same general characteristics as wind waves except for their enormous length, stretching halfway around the world from crest to crest. Their height at sea is so insignificant that the seafarer has no inkling of their presence offshore, but in coastal and estuarine waters their range between high and low water can be considerable. They generate, in turn, their own kind of currents, usually but not always reversing with each tide.

Intermediate in size between the shorter wind waves and the longer tidal waves are the tsunamis, or seismic waves, brought about by disturbances of the sea floor. These are also unnoticeable at sea, but fearsomely destructive when they reach the land.

Thus the boundary at which sea and atmosphere meet is an area of great energy transfer. It is also an area used by man for recreation and practical purposes and so becomes a natural object of curiosity. Its complex and varied movements may at times appear chaotic and unpredictable. But there is a pattern—a grand design—to all this, which forms the basis of this book.

2. *The Simple Wave*

When we talk about ocean waves we are dealing with a whole family of waves that have much in common but result from quite different causes. First, the familiar waves at sea are wind generated. Several different kinds and sizes of wind-generated waves occur, such as ripples, chop, sea, swell, and surf. Isolated freak waves of great size, which are an enormous danger to ships, also occur, at an unexpected moment and from an unexpected quarter. When wind waves leave the deep ocean and approach the shore they change their characteristics to a marked degree. Less familiar, yet also potentially very destructive, are the very much longer tsunamis, or seismic waves, often incorrectly called "tidal waves." Tsunamis are caused by violent earth movements, such as submarine earthquakes or landslides. Meteorological phenomena, such as sudden changes of air pressure, may also bring about dangerous surges.

The tides themselves are also waves, but extremely long and generated by the gravitational forces of the sun and the moon and the rotation of the earth. One result of tidal movement under special circumstances is the tidal bore, a wave that sweeps up some rivers and estuaries as a wall of water. On shallow beaches one may sometimes see waves that are isolated crests widely separated from each other. These are called solitary waves.

Arising from various causes are the internal waves of considerable size, which travel beneath the surface of the sea, not directly visible to the shipboard observer, though of great concern to the submariner. There are also standing waves, or seiches, which do not move progressively but stay in one place. Some standing waves result from tidal action, some occur as a result of waves reflecting from a barrier such as a seawall, and others from a variety of other causes. Finally there are the waves, including bow waves and wakes, produced by the passage of a ship through the sea. The naval architect is particularly concerned about these because they represent a loss of power and must be taken into account in hull design.

Waves at sea usually form a complicated and irregular pattern. The sea surface in a windy area does not consist of a series of orderly waves, all of the same size and shape, with their long crests advancing like a series of troops in parallel lines. A simple wave train such as this is only approximated in the long swell that has moved far out from the storm that produced it. Instead, wave trains of various sizes, often coming from different directions, combine to form a series of hillocks and valleys. In order to understand this pattern and be able to predict seas, it is first of all necessary to consider a simplified wave form and an individual wave train with waves each of the same length and height and each presenting a long, unbroken crest. Once this is understood, it may be used to combine all the varied components that make up the more complex sea that is actually encountered.

For a start it will also be convenient to consider waves in deep water only, since an important change takes place as waves enter shallow water. The three wave shapes that have been used for various theoretical analyses are the sine wave, the trochoid, and the cycloid. Each of these has certain advantages in explaining wave action, although none of them fits the real wave under all circumstances. The cycloid more closely resembles the steep waves of various lengths in the area of a storm, while the sine wave more closely resembles the form of swells, the long waves with rounded tops and relatively low height that run out of the storm area and travel thousands of miles before breaking onshore. Intermediate in form between the sine wave and the cycloid is the trochoidal wave.

A sine wave may be drawn by considering the steady movement of a point on the rim of a wheel. As the wheel rotates, the height of the point above or below a horizontal line through the axle changes. By plotting this height along a line divided into equal parts representing equal intervals of time, a smooth and symmetrical wave curve is formed, with

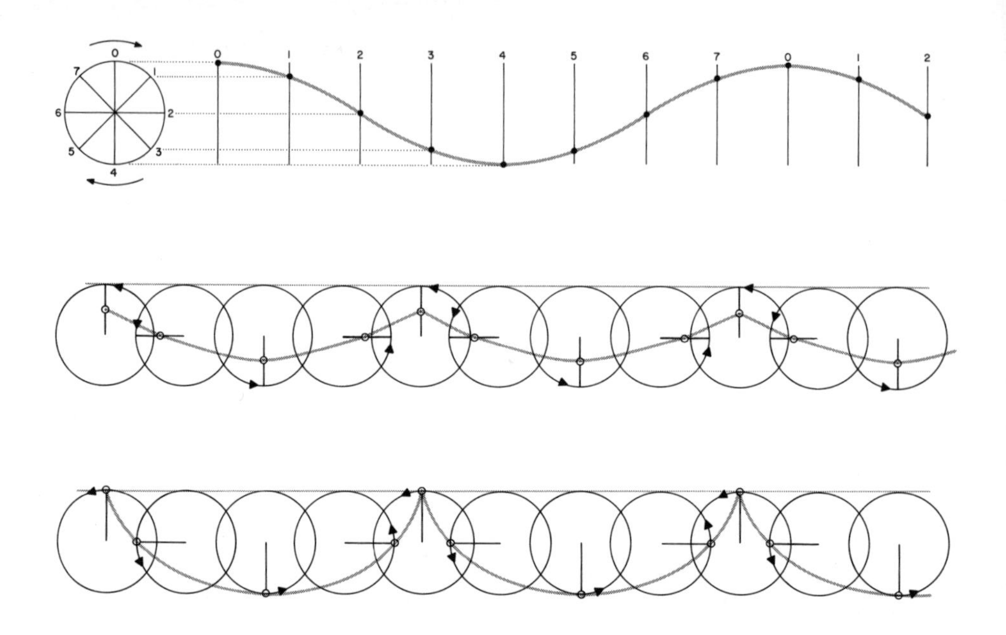

A long low wave, the sine wave, is represented by the first wave form. It can be generated by plotting the vertical height above or below a horizontal line of a point on the rim of a rotating wheel. As the wheel rotates, the height changes. The heights are spaced along the horizontal line at distances proportional to the rate of rotation of the wheel. Figures on the wheel show a complete revolution divided into equal parts. Figures to the right indicate a similar time division along the direction of propagation of the wave.

A wave of appreciable height, the trochoid, is represented by the second wave form. This may be generated by rolling a wheel along the underside of a horizontal line. As the wheel progresses, a fixed point on one of the spokes will trace the wave.

A breaking wave, the cycloid, is the third wave form. It is drawn in a similar manner to the trochoid, except that the point that generates the curve is at the extreme tip of the spoke. (Richard Marra)

the trough having exactly the same shape as the crest, but inverted.

The trochoidal wave may be drawn from the movement of a point on a spoke of wheel rolling along the underside of a horizontal line. As it moves, a curve is traced out by the point. This wave form is somewhat more pointed at the crests and flatter in the troughs than a sine wave. It more closely resembles waves in a sea. If the point on the spoke is moved closer to the rim of the wheel, it will trace out a cycloid, in which the crests are sharply pointed, as in a wave about to break.

In order to understand simple wave motion it is convenient to begin with a sine wave of relatively low height (compared to length). The principal dimensions of a wave are its length, which is the horizontal distance from crest to crest, and its height, which is the vertical distance from crest to trough. For some purposes the term amplitude may also be used. This is the distance between the surface of the sea at rest and either the trough or the crest and is therefore equal to half the wave height. Another property of a wave is its steepness. This is the ratio of the wave height to the wavelength.

An important characteristic of all waves is their period, that is, the time elapsing between the passage of two successive crests at a given place. Although all waves behave in much the same fashion, their periods range between wide

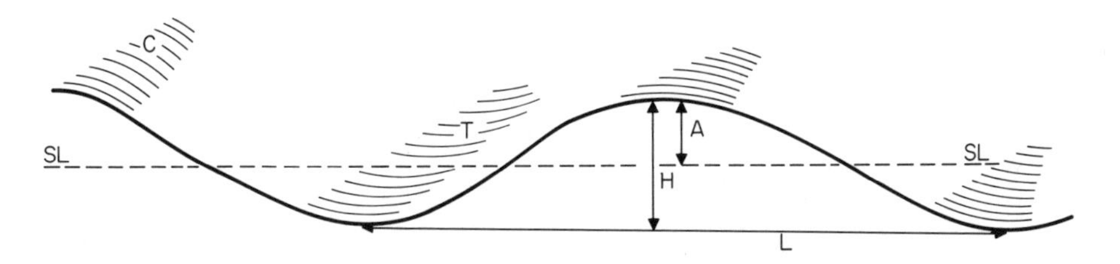

A wave's measure: The principal dimensions of a simple wave are SL, *the level of the water surface at rest;* C, *a long continuous crest;* T, *the trough;* H, *the height; and* L, *the wavelength.* H/L *is the steepness;* A *is the amplitude. (Richard Marra)*

extremes. The period may be a fraction of a second in the case of very small waves or ripples, or as much as 15 seconds in a large storm wave or swell. The period of a tsunami is longer, perhaps 10 to 20 minutes. At the extreme, the principal tidal waves have periods of more than 12 hours.

The period of a wave is proportional to the length from crest to crest so that waves with longer wavelengths have longer periods. The shortest of all waves found at sea are less than one inch in length. These are the tiny cat's-paws, which are the only disturbances of the water surface during light winds of from 1½ to 3 knots. They are called capillary waves.

Capillary waves differ in a peculiar fashion from other wind-

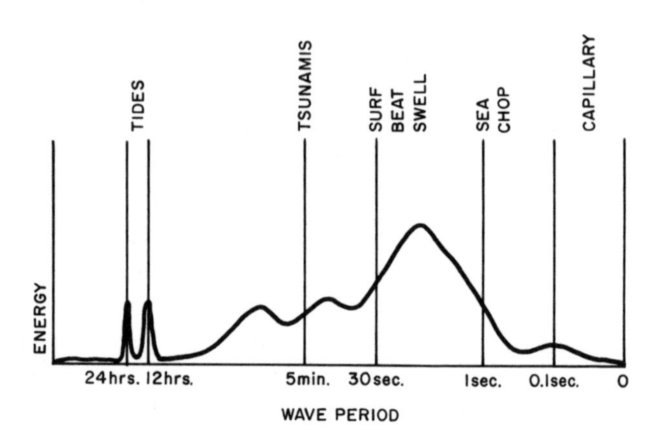

Ocean waves come in all sizes. The smallest is the tiny ripple or capillary wave formed when the wind first disturbs the water. This has a period of one second or less and a wavelength of an inch or so. At the other extreme are the tidal waves with periods from about 12 to 24 hours or more. In between, most of the energy is found in sea and swell, with periods of 7 or 8 seconds up to nearly 1 minute and wavelengths of up to several hundred feet.

induced waves. Instead of growing longer as the wind increases, they grow shorter. This odd behavior arises out of the two forces that propagate the wave. The wind disturbs the water surface and displaces it from the general level of the sea. In the case of ordinary wind waves, the surface is then returned to the general level under the influence of gravity. It is for this reason most wind waves are called gravity waves. In the case of capillary waves, however, the level is restored by the surface tension, or elastic skin, of the water. Another strange feature of the capillary waves is that the crests are rounded and the troughs are pointed in a downward direction, which is exactly opposite to the form of gravity waves, whose crests are pointed and troughs rounded. Still another characteristic of capillary waves is that as soon as the wind stops, the waves subside immediately, unlike gravity waves, which may travel thousands of miles away from the storm areas in which they were generated before losing all of their energy. This accounts for the common observation at sea on a calm day when a light wind begins to stir the water. Suddenly an area of ruffled water will move rapidly across the surface. As the wind passes, the ripples disappear just as suddenly, leaving behind them a sea surface with the original mirror-like appearance.

The behavior of capillary waves and, indeed, of waves in general can be very elegantly demonstrated in a swimming pool. Since the water in a pool is clear and transparent, waves are not easily seen by direct observation. Instead, advantage may be taken of the fact that wave crests refract the light passing through them from above. In an outdoor pool, with the sun fairly high overhead on a bright cloudless day, the crest of even a tiny wave concentrates the light into a brilliant line on the pool bottom. A similar effect is obtained in an indoor pool if a powerful spotlight is used as a light source. The generation of capillary waves may be seen in this way when gentle puffs of wind disturb the surface of the pool. Or, on a calm day, an electric fan may be used. Gravity waves may also be generated by suddenly plunging a short plank of wood beneath the surface with both hands so that the entire length of the plank is simultaneously submerged an inch or so. With the use of this improvised wave tank it is possible to demonstrate without difficulty many of the properties of waves, as they are described in the following chapters.

The speed of a wind wave in deep water is proportional to its period. Waves of long period travel faster than those of shorter period. Thus, a yachtsman wishing to ascertain the speed of a swell or a reasonably simple wave train has merely to time the number of seconds for passage of two successive crests past a free-floating, stationary object and multiply this by 3. The answer is a rough approximation of the wave speed in knots. The speed of a 4-second wave is therefore about 12 knots.

Similarly, to estimate the length of a wave, a rough approximation is obtained by multiplying the square of the period (in seconds) by 5. The result is wavelength in feet. In the case of a 4-second wave traveling at 12 knots, the length is 80 feet.

If a seaman is able to measure the length of a wave, by comparing it with the length of his vessel, for instance, it is a simple matter to determine its speed. He merely multiplies the square root of the length by 1⅓. Thus a 100-foot-long wave has a speed of about 13 knots, a 50-foot wave has a speed of 10 knots, and a 10-foot wave has a speed of 4 knots. Calculations such as these may be useful in determining whether a boat is able to increase speed sufficiently to keep abreast of the crest, or to surf ahead of it.

1. RELATION OF VELOCITY AND LENGTH FOR WAVES OF DIFFERENT PERIODS IN DEEP WATER

Period (seconds)	Velocity (knots)	Length (feet)	Period (seconds)	Velocity (knots)	Length (feet)
2	6	20	14	42	1,003
4	12	82	16	49	1,310
6	18	184	18	55	1,659
8	24	328	20	61	2,046
10	30	512	22	67	2,475
12	36	738	24	73	2,948

Since speed, length, and period are interrelated, they can be calculated from the approximation, wavelength divided by period equals 1.7 times the speed.* From this it can be seen that if a person aboard a ship is able to measure any two of the three—speed, length, and period—he may calculate the remaining one by simple mental arithmetic.

It is apparent that the principal effect when a wave passes by is a movement of the wave form rather than a horizontal passage of water. This is fortunate for the mariner, for if in a 30-knot wave the entire solid water were also to travel at this speed, its impact on a vessel would be catastrophic. A cork floating on the surface will rise and fall with the passage of a wave, but it will remain behind after the wave has passed. Particles of water move as the wave passes but in a not quite closed circular orbit so that they return almost to the point from which they started. For the moment it will be assumed, for the sake of simplicity, that the orbit is completely closed.

The orbit is equal in diameter to the wave height. In the trough, the surface particles are moving toward the oncoming crest. As the crest approaches, the particles begin to move upward along the circumference of their orbits; as the crest passes, they move in the same direction as the crest. As soon as the crest has passed, the particles immediately begin to

---

* Using units of feet, seconds, and knots, the following relations exist between wavelength (L), period (T), and speed (C):

$$C = 3.07T; \quad L = 5.12T^2; \quad C = 1.36\sqrt{L}; \quad L = C^2/1.8; \quad 1.67C = L/T$$

move downward until once more in a trough. This is readily understood when watching a field of corn during a period of gusty winds. Waves, remarkably similar in appearance to ocean waves, progress across the field, but the individual corn stems bend back and forth, returning to their original upright positions.

A familiar experience of divers is that even below the surface waves they find themselves moved back and forth by a surge. The depth to which this will be felt depends upon the height of the wave. Below the surface the water particles move in similar orbits to those at the surface, but of decreasing diameter, until at a depth equal to one wavelength, the orbit is only 1/532 of the wave height. At this depth the particle movement is therefore negligible. At a depth equal to half the wavelength, the orbit diameter is 1/23 of the wave height. Thus, it is readily seen how a submarine may be comfortable at a comparatively small distance below the surface even though violent storm waves are being generated above.

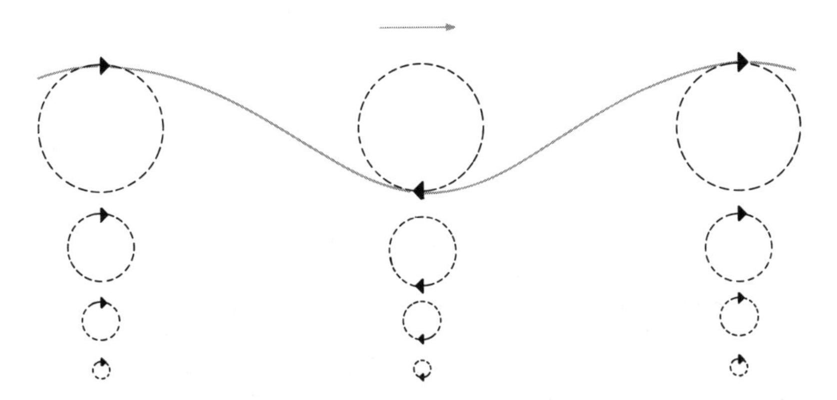

In circular orbit: The dotted lines in the diagram indicate the movements of water particles during the passing of a wave. As the wave passes, the particles do not move with the wave but complete the orbit and return to their starting point. At the crest the particles are moving in the direction of the wave; in the trough they move in the opposite direction. The orbits of particles below the surface become increasingly smaller. At a depth of half of the wavelength the orbital movement is very slight. (Richard Marra)

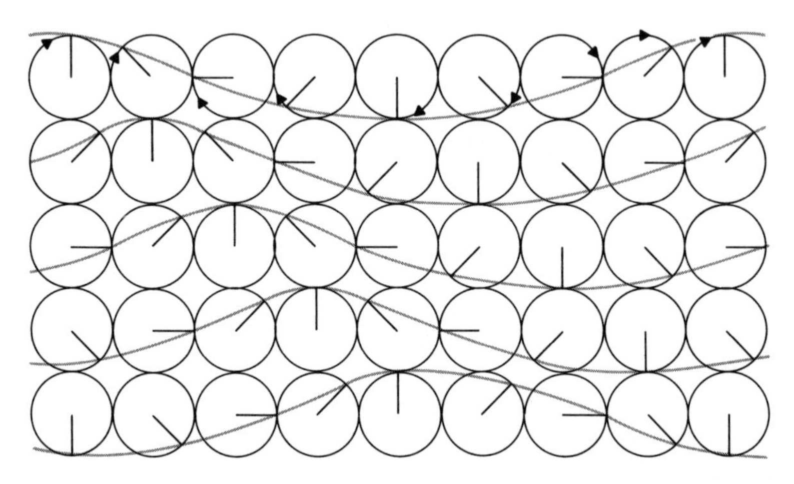

Many circles make a wave. The diagram shows how orbital movements of water particles are related to wave progression. Each row of circles represents the position of a water particle in a wave at intervals of one-eighth period, with the crest moving from left to right. Each column of circles represents the same particle of water at successive intervals of time. The left-hand column, at the wave crest, shows the particle moving forward at the top of its orbit. Beneath it, one-eighth of the wave period later, the particle has moved forward and downward in its orbit, and the back of the wave is falling. Beneath this, again the successive intervals show the particle falling in its orbit, while, at the bottom, it is in the trough. From left to right, each successive column shows the movement of a particle at a different part of the wave. For instance, the middle column shows, at the top, the wave trough. At each interval beneath it, the water particle is moving backward and upward until it forms the crest. (Richard Marra)

The orbital motions of the surface particles at the crest and the trough provide a very interesting partial explanation of the experience that every boatman or yachtsman has, namely that his vessel travels faster with a sea than against a sea even if the sea is a gentle one. The speed of a water particle at the surface as it travels around its orbit increases as the wave height increases. But this orbital speed * becomes smaller as the period and wavelength become greater. For example, a 5-foot-high wave with a 5-second period would have an orbital speed of only about 2 knots at the surface, even though the wave crest itself travels at 15 knots.

If the vessel is moving before a sea, the speed of the water at the crest will add to the speed of the vessel and it will, therefore, stay longer on the crest than in the trough. During

* **Orbital speed (in knots) = 1.9H (feet)/T(seconds)**

this time it is moving faster over the bottom. While in the trough it is traveling slower over the bottom, because of the backward movement of the water. The crest therefore overtakes it faster. The net effect is that it spends more time traveling faster than its normal speed than it does traveling slower. Put more simply, a ship moves faster in a following sea.

Let us return to the 5-second-period, 5-foot wave traveling at 15 knots. If the boat had a speed of 13 knots, the additional 2-knots orbital speed would keep it continuously on the crest traveling at 15 knots, the same speed as the wave, and thus gaining 2 knots over the bottom. Even if the boat did not travel fast enough to remain on the crest, the difference in time spent on the crest than that spent in the trough would be increased, and the boat would, on the whole, travel faster over the bottom than in a calm sea.

If heading into a sea, the vessel will lose speed over the ground while at the crest and gain by the same amount when in the trough. However, this will cause the boat to pass the crest more slowly than the trough, and so the boat will spend more time at the slower speed of the crest. The net result is slower travel over the ground.

Another effect of the opposite motions of the water surface at the trough and the crest is of considerable importance when one vessel is towing another. With a high sea running, it is important that the towing vessel and the towed vessel be in phase with the waves. This means that both must be at the same part of the wave form at the same time. For instance, when a crest rises under the tug, the towrope should be of such a length that a crest is also rising under the tow. Under these conditions both vessels are receiving a similar forward thrust at the same moment as a result of the forward orbital movement of the water. Similarly, when the tug is in the trough, tug and tow are both receiving a similar backward thrust. If, on the other hand, the length of towrope is half a wavelength or 1½ or 2½ or any fractional number of wavelengths, then the tow will be in the trough when the tug rises to a crest. If one is heading into the sea, the result will be that just at the time that the tug rises to a crest and is slowed down by the orbital motion, the tow, in the trough, is speeded up, and the cable goes slack. When the tug drops into the trough and is speeded up, the tow, arriving at the crest, is slowed down, and there is a sudden tautening of the cable. The dangers are readily apparent. And, of course, a similar effect occurs when running before the sea.

The dimensions of simple waves so far considered, length, period, and wave speed, have fairly constant relationships. The dimension of wave height, on the other hand, is mainly dependent upon the strength of the wind and only partially dependent upon the wavelength, speed, and period. There is, nevertheless, an upper limit to the height that a wave can attain in relation to its length. At this height the water in

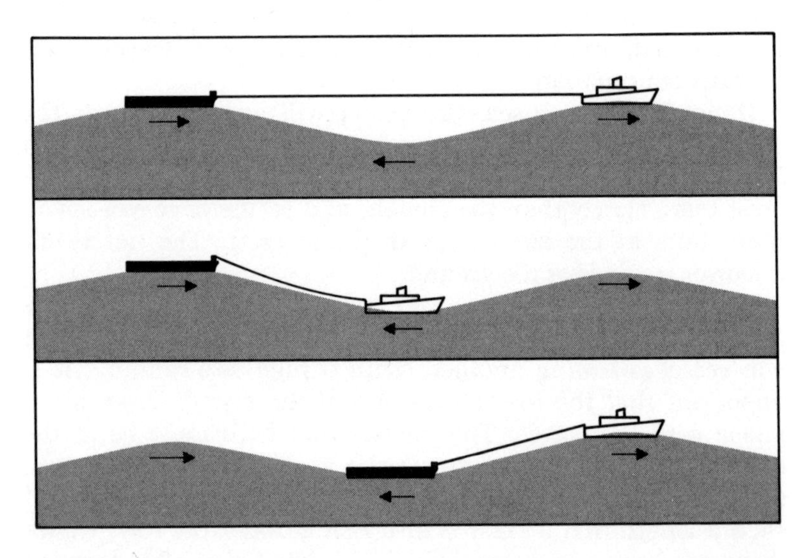

Towing troubles: The waves in these diagrams are moving to the right. The orbital movement of the water particles causes a forward movement of water at the crest and a backward movement in the trough. Thus, in the top diagram a tug and its tow will be subject to the same water movements if both are a whole number of wavelengths apart, since both will be in the trough or at the crest at the same time. If the ships are not a full wavelength apart, then the tug will be in the trough when the tow is at the crest, as in the middle diagram. Here the tug will be slowed down and the tow speeded up, thereby slackening the cable. In the lower picture the ships are subjected to opposite motions, thus tightening the cable.

the wave crest begins to move forward faster than the wave itself. As this happens the water is flung forward ahead of the crest, the crest disintegrates, and the wave breaks. Theoretically, the maximum height is equal to one-seventh of the wavelength, but in actual conditions at sea the height rarely exceeds one-tenth of the length.*

The point at which a wave breaks depends upon the orbital speed of the water particles in relation to the wave speed. Here it must be explained that higher waves do not follow the simple orbital pattern of the ideal wave, which we have so far discussed for the sake of simplicity. As the wave becomes higher, the orbit departs more and more from its circular path, and the particles begin to move faster at the top of the circle than at the bottom. The result is that there is a net forward horizontal movement of the water at the crest. A free-floating cork will be seen not only to move up and down, but also to jog forward at the top of each wave, progressing in the direction of the wave in a series of jerks as each crest passes under it. Another familiar example is that of the boy whose model boat has drifted out of reach on the surface of a pond. He has learned that by tossing a stone into the water just beyond the boat a wave is formed that moves the boat a small distance toward the shore. By repeatedly tossing stones, he eventually brings his toy within reach. Clearly, the wave causes a residual water movement in the direction of its progress.

The increased velocity of water particles at the crest of a wave results from the forward movement of the particle at the

---

* The orbital distance at the surface of a wave is the circumference of a circle whose diameter is equal to the height of the wave. The circumference is therefore πH. This distance is traversed by a particle in the time elapsing between the passage of one crest and its successor, the period of the wave. Thus, the orbital speed is $\pi H/T$. The speed of the wave is $\frac{L}{T}$ so that orbital speed reaches the propagation speed when $\pi H/T = L/T$; $\pi H = L$. From this, the maximum height of a wave would appear to be reached when it is $1/\pi$ of the wavelength, but this does not happen. As the height increases, the upper part of the orbit travels faster than the bottom part and develops speed greater than $\frac{\pi H}{T}$ until, with high waves, it becomes $\frac{7H}{T}$. This equals the propagation speed, $\frac{L}{T}$, when $7H = L$ and the wave breaks.

crest of the wave coinciding with the forward movement of the crest, whereas its backward movement in the trough is in the opposite direction to the wave movement. Consequently, the forward particle movement at the crest continues for a longer period of time than the backward movement in the trough. The net result is that the particle moves forward.

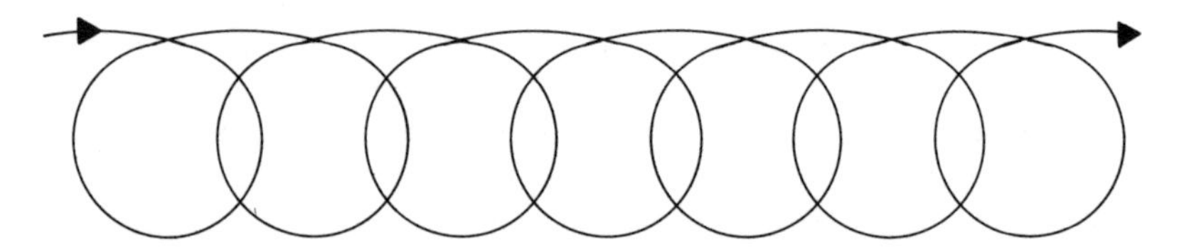

Steep waves move water: Although the figure on page 11 showed the water particles in a wave following a closed circular orbit, this is not completely true in steep waves. The orbit is not fully closed, so that there is a movement in the direction of the wave. This is, however, much slower than the speed of the wave itself. (Richard Marra)

As the wave begins to steepen it changes shape. It departs first from the sine form with a symmetrical crest and trough and becomes trochoidal. In this form the crest is narrower than the trough and projects more above the level of the water at rest than the trough sinks below it. Finally, at its steepest, when height approaches one-seventh of the length, the crest has become pointed, and the wave assumes a shape that is closer to that of a cycloid. At maximum height the angle where fore and back slope meet at the crest is 120 degrees, according to theory.

The ideal wave in deep water has a speed and wavelength that are unrelated to the depth of the water. This is understandable, since, as pointed out earlier, the orbital motion is reduced to $\frac{1}{23}$ of the wave height at a depth of half the wavelength below the surface and to $\frac{1}{532}$ of the wavelength at a depth equal to the length. But when the depth is less than half a wavelength, the orbital motion is appreciably affected by the proximity of the sea floor. As sailors say, the wave begins to feel the bottom. As the depth decreases, there is no longer sufficient room between the surface and the bottom for the water particles to complete their orbits. The orbits become flattened, or squashed, from above, so that they are now ellipses. In very shallow water, the orbits next to the sea floor are reduced to a simple horizontal to-and-fro movement with no vertical rise and fall. The result is that the wave begins to slow down.

In shallow water, though wave speed is less, the period remains the same—the same number of waves pass a given point in shallow water in any given period of time as in deep water. This must obviously be so. If fewer waves passed out of an area in a certain time than entered it, clearly the number of waves in that area would continually and infinitely increase. If more waves passed out than entered, there would soon be no waves in the area. This is noticeably not true. It is also, in fact, comparable to what happens when cars on a highway enter a reduced speed zone. The result of reduced speed does not change the number of cars that leave or enter each minute but it does change the distance between the cars. Just as the cars in a reduced speed zone become crowded together, so do the waves crowd together in the reduced speed of shallow water.

In shallow water, then, the wave period remains the same

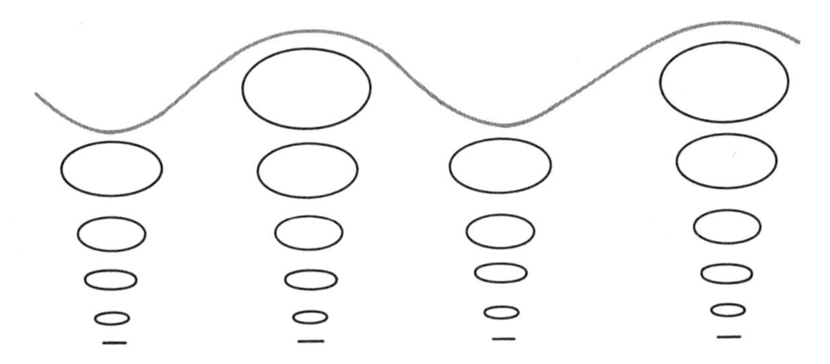

In shallow water there is not sufficient depth for the water particles to complete a circular orbit. The orbits are therefore compressed into oval shapes and, at the very bottom, to horizontal to-and-fro movements. (Richard Marra).

but the speed slackens and the wavelength is shortened. The speed no longer depends on wavelength but on the depth of the water alone. The speed in knots is about 8 times the square root of the depth in fathoms. Thus, in one fathom of water, all waves with a wavelength that is large compared to the depth will travel at a speed of 8 knots. At a depth of 4 fathoms the speed will be about 16 knots.*

2. RELATION OF VELOCITY AND DEPTH OF WATER

Depth of water (fathoms)	1	2	4	8	12	16	25
Velocity of long waves (knots)	8	11	16	20	27	31	40

If one gazes out to sea when waves are breaking on the beach he will notice that well offshore the waves are fairly low and long, but that they become steeper as they approach land, until finally they break. Part of the increase in steepness results from the crowding and the shrinking of wavelengths in shallowing water. Thus, they would become steeper since steepness is the ratio of height to length. But not only does the wavelength decrease, the height itself increases and thus adds further to the steepness.

There is a reason for this increase in height. If it is as-

* A more complete formula for the speed of gravity waves is given by

$$c = \sqrt{\frac{gL}{2\pi} \tanh \frac{2\pi d}{L}}$$

for short waves, where $\frac{d}{L}$ is larger than one-half,

$\tanh \left(\frac{2\pi d}{L}\right)$ approaches 1 and

$$c = \sqrt{\frac{gL}{2\pi}}\,.$$

For long waves, where $\frac{d}{L}$ is smaller than one-half,

$\tanh \frac{2\pi d}{L}$ approaches $\frac{2\pi d}{L}$ and

$$c = \sqrt{gd}.$$

sumed that no energy is lost when the wave slows down and becomes shorter, then the same amount of energy must become crowded into a smaller distance. This causes the wave height to increase, since the energy of a wave is roughly proportional to the square of the wave height. A similar phenomenon also occurs when waves are progressing against a current, when short waves are caused to break as they ride the crest of a long wave and in a tide race, as will be seen in the following chapters.

It is important here to stress the difference between long and short waves. Sometimes these are also called shallow-water (long) and deepwater (short) waves. These terms refer not to the absolute length of the wave nor the depth of the water but to the ratio of wavelength to water depth. The significance of this will be understood when it is pointed out

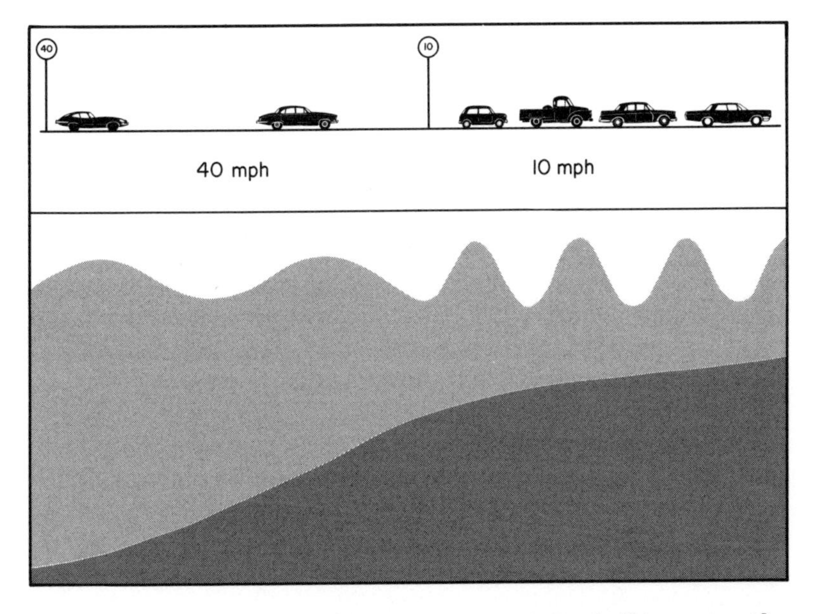

Slowing means growing: When waves move into shallow water they slow down, which causes them to crowd together, thus shortening their wavelengths. This is analogous to the way automobiles crowd together when they go from a fast speed zone to a slow one. When waves crowd together their wave heights increase. (Richard Marra)

that tidal waves and tsunamis act as shallow-water, or long, waves even in the open ocean. This is because their wavelength is more than twice the water depth, and depth, therefore, becomes the controlling factor. By the same reasoning, deepwater, or short, wind waves at sea acquire the character of shallow-water, or long, waves when they approach a beach. Other changes take place when a wave enters shallow water. These will be considered in later chapters when the simplified ideal wave concepts are used to explain the origin of swell and the nature of breakers and surf as they actually exist in nature.

3. *Making Waves*

That winds at sea generate waves of all sizes is evident, but how the energy of air in motion is converted into the moving form of a wave in the ocean is a question that probably occurs to few people other than scientists—even sailors take it for granted. Not surprisingly, since it is not a simple matter to explain, a number of theories have been advanced and discarded. Even today there is no complete agreement and no single theory capable of explaining all the stages of wave development, from the smallest cat's-paw to the larger waves that occur during a storm.

It is fairly easy to comprehend how a wave, once formed, will continue to grow. With a moving wave form already in existence one might expect that a drop in the air pressure just ahead of the crest or an increase in the air pressure behind it would tend to lift the water in front and depress it behind, thus accentuating the forward movement of the crest and increasing the wave height and energy. In addition, any wind movement that exerts frictional drag on the surface, in the direction of the orbital movement of the water at the wave crest, might increase the rate of orbital movement and therefore add energy to the wave. Such mechanisms, however, fail to explain how water in a flat calm is first distorted into the wave form.

When the wind blows across a smooth water surface it does not simply move the water along with it in the way one solid sliding over another tends to exert a frictional drag. Instead, it sets up a kind of oscillation somewhat in the way that the string of a cello is caused to vibrate by drawing a bow across it. Any disturbance of the water surface, in fact, is apt to produce an oscillation or wave, as when a stone is thrown into a pool. The first waves to be formed are capillary waves and small ripples. They can be seen as cat's-paws, which suddenly appear when a light wind blows across a calm surface of water at a speed of more than 2 knots. These initial wave formations probably develop because the wind is not a smooth flow but a turbulent one that carries along with it a series of pressure fluctuations. These fluctuations bring about the distortions of the sea surface, which, under the influence of capillary forces at first and then the force of gravity, become oscillations.

Once a ripple has formed, the interaction between wind and wave form enables the wave to grow. In a general way it may be said that unequal wind stresses and small fluctuations of the atmosphere produce waves and that the waves in turn feed back and modify the wind so as to reinforce the stresses in tune with them. In this way the air stresses develop and increase the amplitudes of the rhythmic movements of the water. It may be compared to the manner in which the movement of a pendulum or a child's swing is increased by applying successive pushes in time with the movement.

An early theory held that the distortion of the wind by the

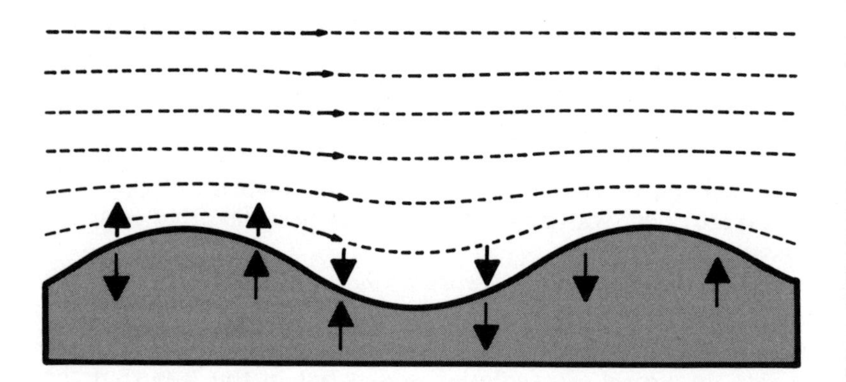

The Bernoulli principle may have a role in developing waves, since the air flow of wind at the crest of a wave becomes restricted, flows faster, and the pressure drops. The reverse happens in the trough. It was thought that this would increase wave height by lifting the crest and pushing down the trough. But as the diagram shows, the decreased pressure at the crest would not only lift the water in front of the crest and so build up the wave, but would also tend to lift the sinking water surface behind the crest, and so dampen the wave.

wave form caused a greater pressure over the trough than over the crest. When a flow of air is concentrated, as over the wave crest, it speeds up and the pressure falls. This is an example of the Bernoulli principle. The boatman is using this principle when he uses a syphon attachment to the dockside water faucet in order to empty the bilge of his boat. In this device the water from the faucet flows through a tube with a constriction. As the water passes through the constriction it is forced to speed up and the pressure drops. The bilge hose is attached to a side tube at the point of the constriction. The consequent drop in pressure in the side tube sucks out the bilge water. The crowding and consequent speeding up of air over the wave crest causes a pressure drop in the same manner. Conversely, the air flow spreads out over the trough and the pressure increases. Thus, the crest would tend to be sucked up and the trough to be depressed.

Unfortunately this nicely symmetrical model does not work. The low pressure at the crest would tend to lift the water not only in front of it, where the water level is rising, but also behind it, where the level is falling. The high pressure at the trough would act to depress the rising water behind the trough as well as to depress the falling water in front of it. The forces thus cancel out their effects upon the wave form. It is clear that the pressure forces must be asymmetrical in order to build up the wave.

Another theory suggested that the air flow blowing over the wave, relatively smooth as it climbs over the back of the crest, tends to push the wave forward. Ahead of the crest the air flow breaks into a turbulent flow and tends to assist the upward movement in front of the crest. This would account for the fact that the front of the wave is steeper than the back. This theory, however, is not fully borne out by actual measurements.

A more recent theory of wave growth is based upon the observation that the rate of flow of the wind is not uniform at all heights above the waves. Because the wind is giving up energy to the waves, it must slow down at levels close to the water. In fact, at wave level, the wind is moving more slowly than the waves themselves and is therefore moving backward

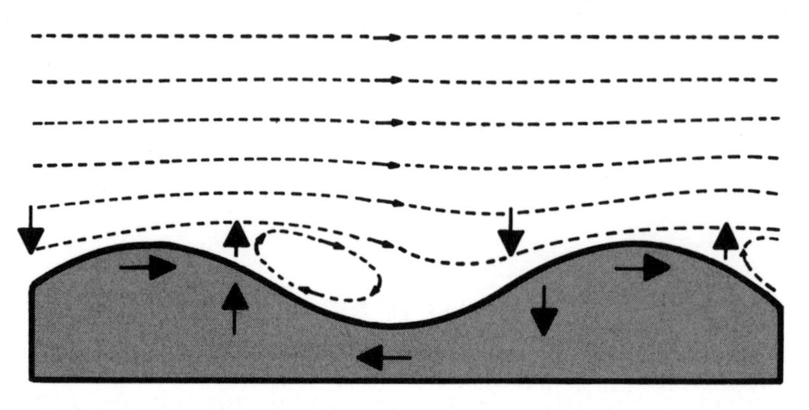

An uneven flow of air caused by the distortion of the air stream by the wave itself was thought to have a role in wave development. The air flowing relatively smoothly against the back of the crest would tend to push the wave forward, while the more turbulent air ahead of the crest would tend to lift the water surface toward the crest. Although this theory explains the formation and movement of waves, it is not well supported by actual measurements.

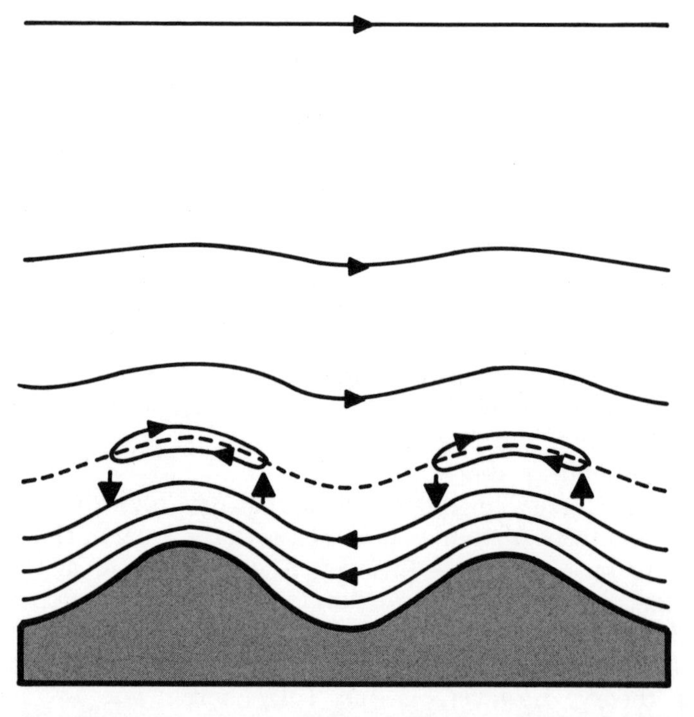

A modern theory of wave development: The transfer of energy from wind to wave slows down the air nearest the water surface. As the wave speed increases, this air may eventually travel more slowly than the wave itself and thus, to an observer traveling with the wave, the wind next to the water will appear to be going backward. An eddy then forms over the wave crest between the layer of air traveling faster than the wave and that traveling slower. This eddy tends to lift the water surface before the crest and depress it behind.

relative to the wave form. Interaction between the more rapid upper levels of air and the slower levels near the sea surface causes an eddy to form above the crest. The eddy is unsymmetrical and produces a decreased air pressure in front of the crest, which helps to build up the crest, while increasing the pressure behind it, which helps to push down the water and so develop the trough.

Although the theories advanced so far may explain the beginning of waves and their further development into larger waves, they are not adequate to explain the largest waves, which may travel at the same speed as the wind or even faster. In such cases, the wind direction relative to the wave is backward not only at sea level but even at some height above the water. Obviously this would tend to dampen the wave rather than increase it. How then are the largest waves formed? The effect of wind friction upon the water particles in orbit at the crest of the wave could be a partial explanation, since the orbital velocity may still be less than the wind speed, even though the wave velocity is faster. The most plausible theory, however, suggests that the shorter waves grow faster than the longer waves, reach their critical height, and break. As they break, their momentum is added to that of the larger waves upon whose slopes they are riding. Thus, the longer waves continue to grow as more short waves are being developed.

One explanation for the slower growth and breaking of the longer waves is that considerably more energy is required to bring about changes in them. But there is an additional reason. As the longer, faster waves move forward, they ride through the slower, shorter waves. As the shorter waves are overtaken they ride backward and upward upon the advancing slope of the following long wave. At the crest of the long wave the orbital movement is forward. This adverse current slows down the movement of the shorter waves as they pass back. It was shown in Chapter 2 that the effect of slowing waves down is to crowd them together. Since the same amount of short-wave energy is crowded into a shorter distance, the wave height is increased accordingly, and if the critical height is reached, the short wave breaks. In the trough, of course, the opposite occurs. The shorter waves begin to separate, become longer, and so less steep. Under these conditions they do not break.

Here it may be convenient to digress in order to point out that the slowing down and consequent increased height of waves when they encounter an adverse current are not only responsible for the manner in which smaller waves contribute to the development of larger waves, but are also the cause of dangerous seas in a number of quite different situations. Near the coast, for instance, especially the south coast of England, short, steep breaking waves may develop in a localized area or an otherwise harmless sea. In this case the adverse current is the tidal stream and the disturbance is known as a tide rip. Another example familiar to yachtsmen is the steep and frequently violent sea that develops in strong ocean currents when waves encounter the opposing flow of the current, such as the Florida current running through the Straits of Florida. A similar situation also develops in inlets and estuaries where outgoing currents are faced by strong swell entering from the ocean. In each case the local development of a dangerously steep and breaking sea results from the slowing down of the wave by an opposing current. Any adverse current in excess of one-quarter of the wave speed will cause it to break.

It is reasonable to conclude that if the adverse current velocity were to become equal to that of the wave velocity, the wave would no longer progress but would remain stationary. There are situations in which this actually occurs. A good example is that of a weir or waterfall. Waves are developed at the edge of the fall and for a distance upstream. It may be observed, however, that the waves are not moving. The surface of the water has a series of stationary wave crests lined up more or less parallel to the edge of the fall. The reason is that the water current is equal to the speed of the wave and so prevents any progress of the wave. In this case, however, the waves do not crowd together, since they were formed in the current and did not enter it from non-flowing water. A similar stationary wave develops upstream of a navigational buoy or a projecting rock in a stream.

The dramatic effects of a strong current running contrary to wind and wave may be experienced vicariously by reading the various published accounts of the unusual accidents that overtook the 39-foot ocean-racing centerboard yawl *Doubloon* in 1964. With her experienced skipper, Jet Byars, and three crew members she left St. Augustine, Florida, early in May on what should have been a routine sail to Morehead City, North Carolina, on her way to Rhode Island, where she was to participate in the Bermuda Race. The course would take her into the axis of the Gulf Stream, running in a roughly northeasterly direction. The next day she encountered very large swells. Soon afterward the wind shifted to the north and increased rapidly to 50 knots, with gusts of 70 knots. From this direction the wind and waves were running in a direction opposite to that of the Stream. By the third morning the seas were running up to 25 feet high and were unusually steep and breaking. Shortly after midnight, with only the wheelman on deck, the crew members below deck heard a tremendous crash. They were hurled about the cabin as the ship rolled completely over, through 360 degrees, losing the mizzen mast, both booms, and most of the mainmast. The helmsman was washed overboard but saved by his safety line. The increased stability resulting from being dismasted might have been expected to prevent further troubles, but the yawl was again rolled over through another complete revolution. In spite of this rare experience *Doubloon* was repaired in time to participate in the Bermuda Race, taking second place in her class. This is a great testimony to the strength of a modern ocean-racing yacht. It also exemplifies the way in which big seas may become steepened as a result of adverse currents to the point of extreme danger.

After the wind has been blowing for a while, the shorter waves reach a critical height and break, as explained in Chapter 2. The longer waves continue to grow and so begin to predominate until maximum heights and lengths are reached. The maximum size depends upon the strength of the wind; the distance of open water over which the wind has been blowing, known as the fetch; and the length of time during which the wind has been blowing, its duration.

The maximum height and length of a wave are determined by the strength of the wind, assuming that there is no interference from waves generated elsewhere and moving into the area. These dimensions are only reached when the wave has been acted upon by the wind for a certain minimum length of time. For any lesser duration, the wavelength and wave height will be less. Thus, a 20-knot wind is able to develop a 9-foot-high wave only after the wave has been exposed to its influence for at least 55 hours. At this time the sea is fully developed and increased exposure to the wind will not materially increase its size. On the other hand, an exposure to this wind for a period of 30 hours will generate waves of 8-foot maxi-

Half hidden behind steep waves as she rounds the offshore mark during ocean races in the Florida Straits, this yacht experiences the effect of an adverse current upon the waves. As the waves travel into the path of the 4- or 5-knot current, the waves steepen and tend to break. One may roughly gauge the location of the maximum current when sailing between Florida and the Bahamas by noting the change in wave height.

mum and an exposure of 10 hours will generate waves of no more than 6 feet.

The role of fetch in developing wave size is best understood when it is realized that fetch is essentially a limitation of the duration. Thus, fetch can be the distance that the wave has traveled in the wind area since it was first developed. Or it can be the distance the wave has traveled from a weather shore. Near the shore, where waves are first generated, they have only begun to be exposed to the wind and are, therefore, small. As they progress out to sea their duration of exposure to the wind increases and their wave size increases. Thus as the fetch, or distance offshore, increases, so does the duration of exposure. When the fetch is great enough to provide the duration of exposure needed for maximum size, then no further growth takes place with increased fetch. In the case of a 30-knot wind, a fetch of 800 miles is equivalent to a duration of 60 hours and a fetch of 60 miles is equivalent to a duration of 9 hours.

A concrete example may be taken from the Straits of Florida where heavy yacht traffic occurs between Miami and Gun Cay, Bahamas. The distance between soundings is a little more than 40 nautical miles. A 20-knot easterly wind would need a fetch of several hundred miles in order to develop maximum seas. With a fetch of 40 miles, however, the maximum sea at the leeward side of the Straits is 5 feet and this will be reached after a duration of 7 hours. Greater duration would not increase wave size over this fetch, but shorter duration would produce a smaller sea.

Another example, based upon measurements made between the coasts of Ireland and Lancashire during WSW winds of 36 knots, shows that waves of 5–6-second period reach their maximum height over a 40-mile fetch and 6–7-second period waves require a 50-mile fetch to reach their maximum height. Longer period waves, 7–11 seconds, however, cannot reach maximum height because the distance between the two coasts, 110 miles, is insufficient. To sum up, over short distances fetch limits wave size, while over great distances duration is the limiting factor.

The height of a wave is dependent not only on the

strength, fetch, and duration of the wind, but also upon the relative temperature of the air and the sea. Other things being equal, a decrease of air temperature to 10 degrees less than the sea temperature will increase the height of waves developed by 25 percent, unless they have already attained their maximum.

The period of a wave is similarly dependent upon the fetch, duration, and velocity of the wind. The period of the average wave in seconds is roughly one-quarter of the wind speed in knots. Thus, a 30-knot wind after maximum duration will develop a wave of 7 seconds. Longer and shorter waves will be developed, of course. The ranges and maxima of wave periods and heights for different winds will be discussed in more detail in the following chapter.

As the sea develops under the influence of the wind, the longer waves begin to travel outward from the storm and eventually outstrip the slower, smaller waves. Not only do the smaller waves travel more slowly, they also lose their energy more quickly. The larger ones, which retain their great energy for a considerable time, are called swell. Their appearance is quite different from the steep waves in a sea. They are more uniform, their crests are longer, smoother, and more rounded, and although their height and steepness decrease as they travel away from the storm, they maintain their identity for thousands of miles.

A train of waves leaving the area of a storm, in the form of swell, does not travel at the same speed as that of the individual wave. This may sound impossible at first, but a little experiment will show it to be so, and further consideration will explain why this happens. Suppose one throws a pebble into a pool. Waves are generated and move out in all directions. The wave crests travel outward as circles. If attention is concentrated upon the movement of a single crest it will be seen that the waves are traveling at a certain speed. If the eye is then focused upon the middle of the group of waves it will be seen that the group is traveling more slowly, at half the speed of the individual waves. Further observation will also show that the front wave does not continue long, but dies and is replaced by one from behind it and that a small wave is continually appearing at the rear and growing to full size as it travels through the group.

The reason that wave groups move at half the speed of the individual wave lies in the manner in which energy is transferred. Only half of the wave's energy actually moves with the wave. Half of the energy, kinetic energy, the kind of energy that exists in a spinning flywheel, is bound up in the orbital movements of particles. Since the orbits of the particle movement remain in the same location, rather than moving along with the wave, kinetic energy remains behind as the wave passes by. However, the remaining half of the energy, potential energy, is the energy that is potentially available by virtue of the fact that the crest has been displaced from the level of the sea at rest. This is the kind of energy stored in a lifted weight. When support is removed from the weight, it drops and releases the energy. In restoring the level of the water to its original level, potential energy is released. Potential energy thus travels with the wave itself.

A simple wave train illustrates this situation. The last wave crest (Number 9 on page 32) passes forward, carrying with it the potential energy. It leaves behind orbiting particles of water, which continue, but with only half the original energy, and therefore this wave is now only half the size. The wave that passes forward carries only half of its original energy, but it picks up the half energy left behind by the wave ahead, which has just moved forward, and so maintains its size. This process takes place along the whole group until the first wave is reached.

The front wave moves forward with only half its energy. Since it enters still water that has no energy, it loses height (Number 1 on page 32). At the next phase, as it moves forward it now carries only one-quarter of its original energy. In this way the whole group moves forward but with dying waves both in front and behind. The group thus travels at half wave speed. To an observer moving with the group, it appears as if small waves appear at the rear, grow to maximum size, travel through the group, and then, as they move out ahead, diminish and die.

So far only simple wave trains have been considered. The

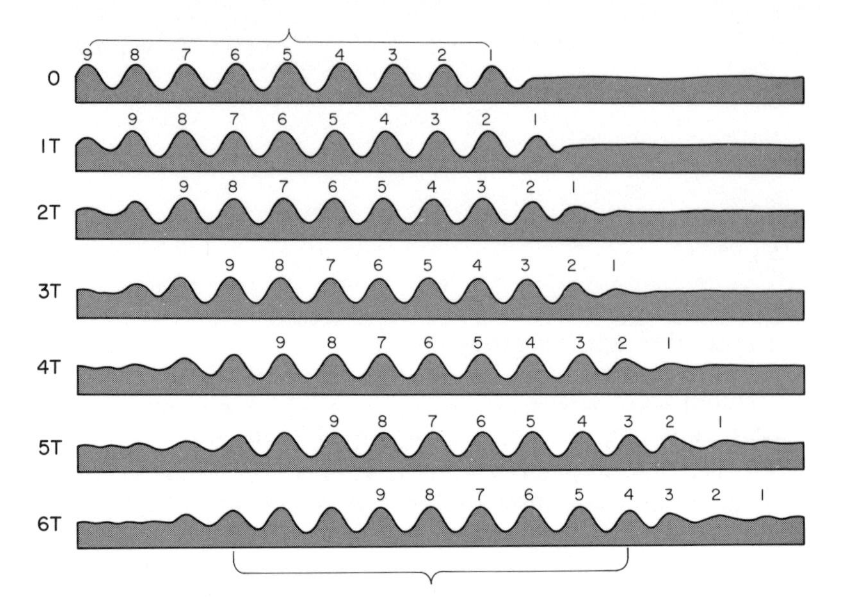

How fast does a wave train travel? Because the waves die out in front and new ones form behind, a train of waves travels at half the speed of a single wave crest. The diagram shows the advance of a train of waves numbered from 1 to 9. The successive positions correspond to intervals of one to six wave periods. Crest 1, advancing, gradually loses its height and is replaced by wave 2. Behind the wave train, as wave 9 moves forward, it leaves part of its energy behind. The middle wave of the train at the beginning is wave 5. At the end of the time it is wave 8. Thus, while each of the waves has traveled a distance of 6 wavelengths, the train itself has traveled only 3 wavelengths.

winds at sea, of course, develop waves of many different periods and heights, and the effect of the interaction of many such waves, moving in the same direction, is to cause a very complicated wave profile. This can be simply illustrated, however, by taking two wave trains of slightly different wavelength running in the same direction. When the troughs and crests coincide, they reinforce each other, so that the waves will become higher than in either train alone. When they are out of phase, that is, when the trough of one coincides with the crest of the other, the resultant wave will be flattened.

Many seamen have a strong belief that every seventh wave will be an extra high one. Others argue that this is true only of the ninth or twelfth wave. There is some basis for this when we consider the preceding paragraph. Because of the interference between two wave trains of different wavelengths, waves may advance in groups of higher waves alternating with groups of much lower waves. The number of waves passing between two successive groups of highest waves obviously depends upon the wavelengths of the different interacting trains so that there can be no magic invariable figure such as

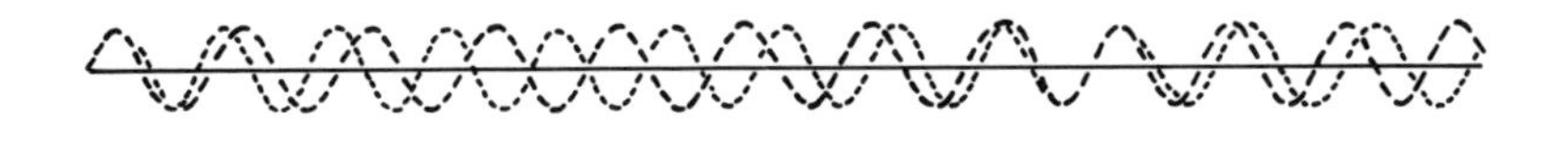

The myth of the twelfth wave: Two wave trains of slightly different wavelength can combine to form alternate groups of larger and smaller waves. When the peaks coincide, the waves become higher. When the troughs of one train coincide with the peaks of the other, the waves will be lower. Thus, it may sometimes appear that every twelfth or seventh or ninth wave is an especially high one.

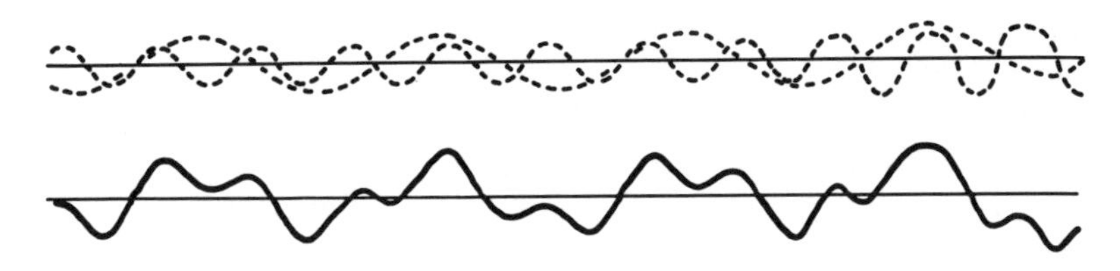

A lumpy sea may form when waves of considerably different wavelengths combine.

seven. On the other hand, the tendency to come in groups, as a result of interference, is commonly experienced, and the careful seaman who desires to come about in a heavy sea will usually measure the time that elapses between the low groups so as to maneuver under the most favorable conditions.

The more complicated conditions in a storm at sea, with waves of different lengths coming from different directions, cannot be described or forecast without resorting to statistical methods and spectral analysis. Nevertheless, the concept of simple wave trains developed in the preceding pages forms a useful basis toward the understanding of more complicated situations.

4. The Real Sea

Although there are many obvious reasons for developing a method of sea state prediction, the greatest effort in this direction was not made until World War II, in response to the need to choose suitable conditions for military landings on the Normandy beaches and elsewhere. Nevertheless many other activities benefit by foreknowledge of wave conditions. The use of sonar is subject to interference from wave action. Radar echoes from the sea surface cause "sea return," or clutter, which masks the echoes from navigational buoys or other vessels. Offshore drilling and harbor construction projects are vulnerable to heavy seas and accordingly benefit by advance warning of dangerous conditions. Knowing the size of waves to be expected, submarine commanders are able to plan ahead to remain submerged at an appropriate depth, below the influence of wave movement.

Heavy seas not only impose drag, or friction, which slows the vessel, but also require deliberate speed reductions in order to ease the strain upon the hull, upon the passengers, and upon the cargo stowage. Head seas are the most obstructive, while following seas, except in rare cases, are less effective in forcing a reduction of speed. Radar efficiency is also reduced by reflection from waves to the extent that upwind search is difficult in sea state 3. The effect of "noise" generated by wave action upon the functioning of sonar also increases with sea state.

The development of sea state forecasting has progressed to the point where the U.S. Navy is now able to provide a routing service to both naval and commercial vessels. This provides continuous information on sea states that may be expected in various areas as much as three days ahead. Ships receiving this service are able to change their routes so as to avoid conditions that would entail a reduction in speed and an increase in fuel consumption. The cost of this service is trivial compared to the savings that are effected, even though the actual distance traveled may be more than a direct great circle route. The average time saved is about 10 percent.

Before it is possible to predict sea conditions, it is first necessary to analyze stormy seas and to develop a numerical language that will realistically indicate their nature and strength. Following this, a relation must be established between actual records of sea states in the past and the conditions of wind strength, fetch, and duration that caused them. In Chapter 3 simple wave trains and the interaction of two trains of different sized waves traveling in the same direction were considered. In a stormy sea, however, the waves that eventually come together differ not only over a wide range of wave heights, lengths, and velocities, but they may also advance from several different directions. The waves that originate in any given place and at any given time travel in the general direction of the prevailing wind, but in other parts of the storm area the wind may blow in different directions, and at any one location the wind may change direction. The result is that seas in a storm may be the result of interacting waves moving in several directions. When this happens the long crests shorten and disappear into a confused surface characterized by a combination of short crests, hillocks, and hollows of different sizes. When the crests coincide there will be a short crest. Where the troughs coincide there will be a short trough. It has been well said that the appearance of such a sea is so chaotic that the same configuration is never repeated exactly.

An excellent example of the way in which cross seas may develop is given by the tropical cyclone. Longer waves move out from the central part of the storm area at a speed faster than the forward progress of the storm. As they move farther out they enter a region in which the sea surface is exposed to winds from a different direction, perhaps as much as 90 degrees apart. This may cause a very confused sea. Also, since the cyclone itself is moving and continually entering a new area of the ocean, winds in that area may change with great rapidity, developing new seas in a different direction from the old seas with which they interact.

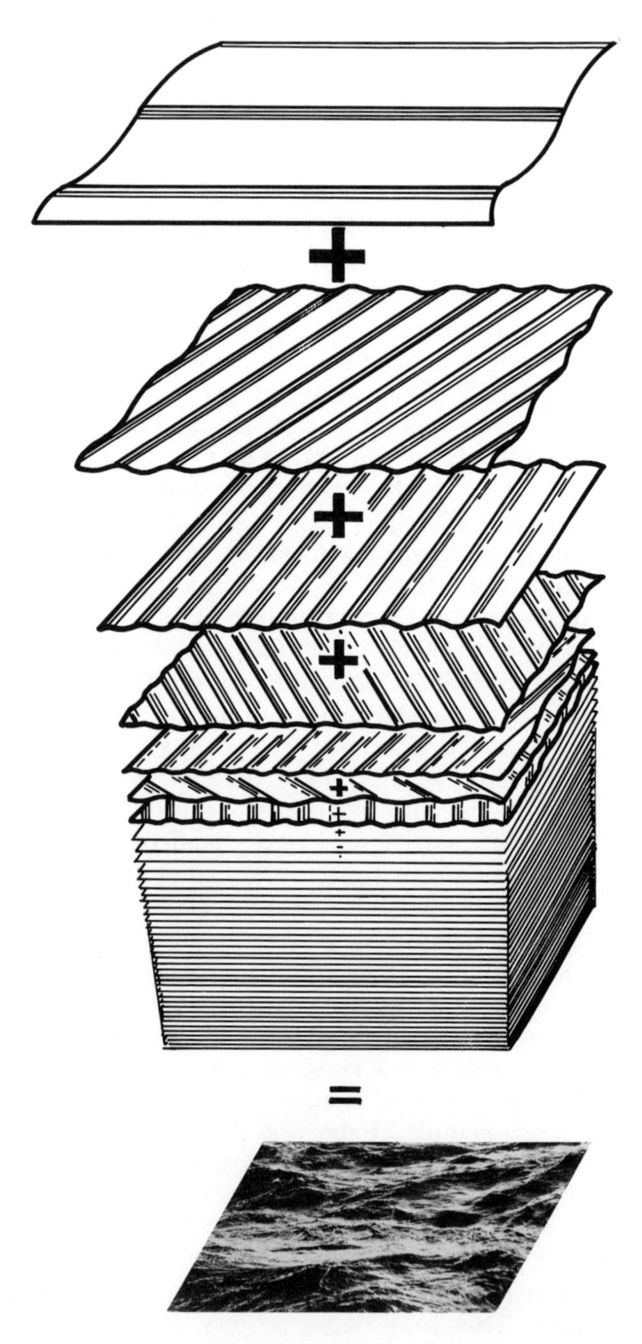

A storm at sea develops a highly irregular pattern of interacting waves of various lengths and heights, frequently moving in different directions. The photograph at the bottom illustrates the surface of a sea, while above are shown the various elements that combine to make the sea. (Richard Marra)

When one looks at the sea in a storm, it is difficult to see how all the wave trains can be sorted out and some kind of accurate description given to it. In fact, it begins to appear as if the theoretical approach of previous chapters, dealing with simple wave trains, may be very valuable in other respects but too helpful in making quantitative statements about the chaotic confusion of the real sea. Since an attempt to forecast the kind of sea that one might expect from any given set of weather conditions must begin with a description of the complex dimensions of the sea, it is necessary to seek additional considerations and criteria.

For many years the solution to this problem was based upon visual observation. A simple, practical approach to the relation between wind and sea was originally introduced in 1805 by Sir Francis Beaufort. This approach relates actual wave conditions to a numerical scale of wind speeds. The wind scale was introduced during the days of sailing ships, and its original purpose was not to estimate the kind and magnitude of sea to be expected from a wind of given speed, but rather the opposite. The skipper of a sailing vessel gauged the strength of winds from the appearance of the sea and so could better judge when to take in or add sail.

When the wind first starts to blow, the sea begins to develop and continues to do so for a certain period of time; after a while, when it is fully developed, the sea no longer increases unless the wind changes. When a sea is fully developed there is a fairly good correspondence between the wind force and the nature of the sea. With the disappearance of commercial sailing ships the Beaufort scale would have lost its value had it not been for its use in forecasting the fully developed sea that may be expected from a given wind force, measured by anemometers on the vessel or derived from meteorological forecasts. In more recent years, however, wave research has resulted in an entirely different method of forecasting, which is much more accurate and which provides a quantitative statement of the nature of the sea rather than a visual description. As a result, the reporting of sea state by the Beaufort Scale is no longer continued by the international weather service. The scale is still of value to yachtsmen, however, since

3. MODERN BEAUFORT SCALE

WIND			INTERNATIONAL	
Beaufort number	Knots	Nautical term	Form and height of waves in feet	Code
0	under 1	Calm		
1	1–3	Light air	Calm, glassy 0	0
2	4–6	Light breeze	Rippled, 0–1	1
3	7–10	Gentle breeze	Smooth, 1–2	2
4	11–16	Moderate breeze	Slight, 2–4	3
5	17–21	Fresh breeze	Moderate, 4–8	4
6	22–27	Strong breeze	Rough, 8–13	5

(Continued on

it gives a means of estimating wind force in the absence of reliable wind gauges. A modern amendment of this scale is shown in Table 3 and the types of seas are illustrated by photographs.

The modern system of wave forecasting relies upon statistics for a method of describing seas developed by known winds. Each individual component wave that contributes to a fully developed sea—that is, the maximum sea for the conditions of wind strength, fetch, and duration—still behaves according to the theory of the simple wave. In order to describe a fully developed sea it is necessary to measure it and break it down into these individual wave components, each with its characteristic period and height. The measurement of seas has been greatly improved by modern wave recorders. Such recorders may be on lightships or on vessels traveling across the oceans. They measure the changing height of the water as the larger and smaller crests pass and also the time

WIND EFFECTS

Effects observed at sea	Effects observed on land
Sea like mirror.	Calm; smoke rises vertically.
Ripples with appearance of scales; no foam crests.	Smoke drift indicates wind direction; vanes do not move.
Small wavelets; crests of glassy appearance, not breaking.	Wind felt on face; leaves rustle; vanes begin to move.
Large wavelets; crests begin to break; scattered whitecaps.	Leaves, small twigs in constant motion; light flags extended.
Small waves, becoming longer; numerous whitecaps.	Dust, leaves, and loose paper raised up; small branches move.
Moderate waves, taking longer form; many whitecaps; some spray.	Small trees in leaf begin to sway.
Larger waves forming; whitecaps everywhere; more spray.	Larger branches of trees in motion; whistling heard in wires.

pages 40 and 41.)

intervals between them. The complicated wave record is analyzed in such a way that the waves are separated by their wavelengths or periods and the average height for each different wavelength or period is measured.The energy for each period is proportional to the square of the wave height. The final result is a graph in which the wave energy is plotted against the wave periods in a manner analogous to a light spectrum. The actual analysis may be performed electronically. As the fluctuating wave heights from the continuous measurement record are fed into the analyzer, they are separated by means of a filter that responds to a particular period. Only wave heights that are recorded at the time interval of this period pass through the filter, in somewhat the same way that a radio tunes in only one wavelength at a time, or the way a color filter separates light of a specific wavelength or color. When this procedure is repeated for a whole range of wavelengths the spectrum can then be completed.

3. MODERN BEAUFORT SCALE (Continued)

WIND			INTERNATIONAL	
Beau-fort number	Knots	Nautical term	Form and height of waves in feet	Code
7	28–33	Moderate gale		6
8	34–40	Fresh gale	Very rough, 13-20	
9	41–47	Strong gale		
10	48–55	Whole gale	High, 20–30	7
11	56–63	Storm	Very high, 30–45	8
12 13 14 15 16 17	64–71 72–80 81–89 90–99 100–108 109–118	Hurri-cane	Phenomenal over 45	9

Since January 1955, weather-map symbols have been based on

The average wave height recorded for each individual wavelength or period, along with the relative predominance of each period, is used to calculate the energy for that wave period. It will be seen that for any particular sea condition there is one wave period wherein lies the maximum energy. Typical wave spectra in a fully developed sea can thus be obtained for each of a whole range of known wind conditions. This allows a forecast to be made when similar wind conditions are met or expected in the future.

In the wave spectrum diagram the wave period is plotted against the wave energy for several different wind velocities.

WIND EFFECTS

Effects observed at sea	Effects observed on land
Sea heaps up; white foam from breaking waves begins to be blown in streaks.	Whole trees in motion; resistance felt in walking against wind.
Moderately high waves of greater length; edges of crests begin to break into spindrift; foam is blown in well-marked streaks.	Twigs and small branches broken off trees; progress generally impeded.
High waves; sea begins to roll; dense streaks of foam; spray may reduce visibility.	Slight structural damage occurs; slate blown from roofs.
Very high waves with overhanging crests; sea takes white appearance as foam is blown in very dense streaks; rolling is heavy and visibility reduced.	Seldom experienced on land; trees broken or uprooted; considerable structural damage occurs.
Exceptionally high waves; sea covered with white foam patches; visibility still more reduced.	
Air filled with foam; sea completely white with driving spray, visibility greatly reduced.	Very rarely experienced on land; usually accompanied by widespread damage.

wind speed in knots, at five knot intervals, rather than upon Beaufort number.

In each curve the area below the curve is a measure of the total energy in the sea. For a 20-knot wind most of the energy is distributed over a small range of wave periods lying between 6 and 9 seconds. When the 30-knot wind is considered the periods and heights of the waves increase, and the greater part of the energy is in waves of periods between 10 and 20 seconds with a maximum at 15 seconds.

For a determination of the fully developed sea it is necessary to know the fetch, duration, and strength of the wind, as explained in Chapter 3. These conditions may be ascertained today from meteorological reports and forecasts, which are

Beaufort force 0. See table on pages 38–39 for explanation.

Beaufort force 1. See table on pages 38–39 for explanation.

Beaufort force 2. See table on pages 38–39 for explanation.

Beaufort force 3. See table on pages 38–39 for explanation.

Beaufort force 4. See table on pages 38–39 for explanation.

Beaufort force 5. See table on pages 38–39 for explanation. (All photographs courtesy Canadian Meteorological Service)

Beaufort force 6. *See table on pages 38–39 for explanation.*

Beaufort force 7. *See table on pages 40–41 for explanation.*

Beaufort force 8. *See table on pages 40–41 for explanation.*

Beaufort force 9. See table on pages 40–41 for explanation.

Beaufort force 10. See table on pages 40–41 for explanation. (All photographs courtesy Canadian Meteorological Service)

the basis for predicting the probable development, intensity, and movements of storm centers. From tables and graphs, this information, in turn, will provide information as to wind direction and sea conditions to be expected at the mariner's locality.

In using wave spectra for forecasting it is necessary to give a simple idea of the height and length of waves that may be developed, since the mariner would like to know the height of a typical or average wave rather than be concerned with the entire wave spectrum. For this purpose various statistical heights are used to describe the sea. Commonly used is the significant height, which is the average of the highest one-third of the waves. It turns out that the significant height

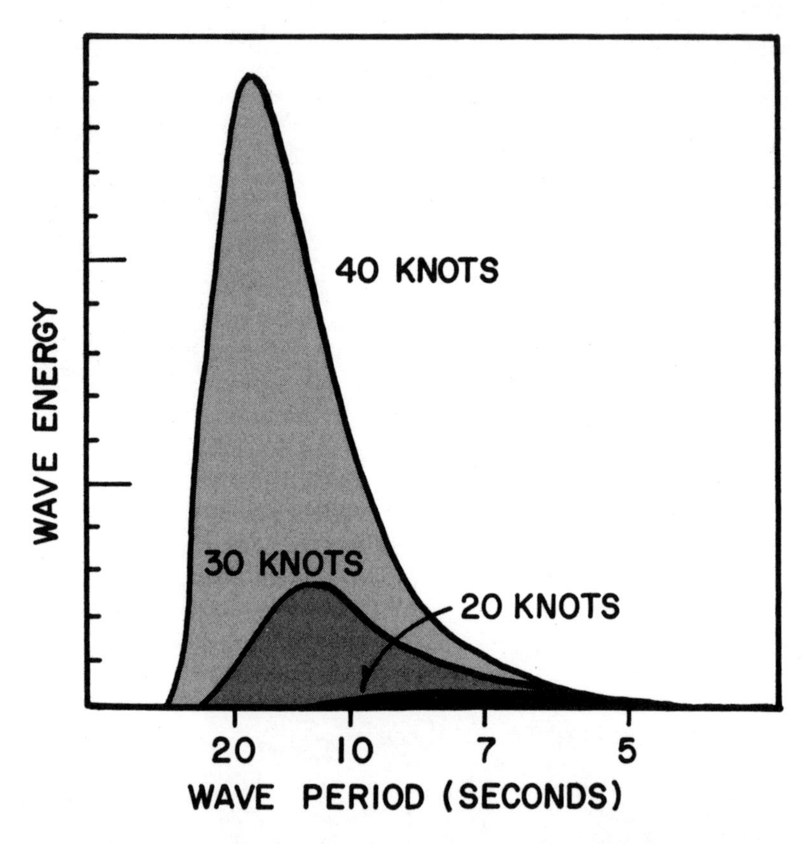

A comparison of fully developed seas caused by winds of 20, 30, and 40 knots shows that the greater the wind the longer the period or length of the waves in which most of the energy is concentrated. In 20-knot winds, the diagram shows that waves range from 5- to 10-second periods, with maximum energy at 7 seconds. Forty-knot winds concentrate energy in a narrow zone of waves of approximately 15-second periods, but with very small amounts due to wave periods above 20 seconds and below 5 seconds. The effects of a 30-knot wind are intermediate. (After Pierson, Neumann and James)

approaches very closely what would appear to be to an experienced seaman the average height. Also used is the average of the highest one-tenth of the waves.

Although the average wave height for a fully developed sea derived from a 20-knot wind is 5 feet, the significant height is 8 feet and the average height of the highest 10 percent is 10 feet (Table 4). With this broader range of information, a fully developed sea can be better described than when discussing wave periods and heights for a single wave. The accompanying table and diagrams show the relationships that have been developed from many observations.

The diagram is used to forecast the significant height and significant period from the fetch, duration, and speed of the wind. The user finds the wind speed on the vertical scale

4. WAVE CONDITIONS IN FULLY DEVELOPED SEAS

Wind Velocity in knots	Length of fetch (nautical miles)	Duration (hours)	Average height (feet)	H_3 significant height	H_{10} Average of the highest 10% (ft)	Period where most of energy is concentrated (sec)
10	10	2.4	0.9	1.4	1.8	4
15	34	6	2.5	3.5	5	6
20	75	10	5	8	10	8
25	160	16	9	14	18	10
30	280	23	14	22	28	12
40	710	42	28	44	57	16
50	1420	69	48	78	99	20

to the left. He then follows this horizontally across the chart to the right until it intersects either the vertical lines of the fetch, which is read off from the horizontal scale, or the duration on the sloping lines, whichever comes first. At this point the height and period are read off from the heavier curves. The full curves represent height and the broken ones period. For instance, in the Straits of Florida between Miami and Gun Cay there is a fetch of slightly more than 40 miles. With an easterly or westerly wind of 20 knots that has been blowing for a day or more the following procedure would be followed. The 20-knot wind with a fetch of 40 miles produces an intersection at the 6-hour duration curve. If the duration is more than 6 hours, the effect is still limited by the fetch and the intersection provides the forecast, no matter how long the wind has blown. At this intersection the chart shows a wave height of 5 feet and a wave period of between 5 and 6 seconds. If the duration was less than 6 hours, for instance 4 hours, then the duration would limit the effect, and the intersection of 30 knots and 4 hours must be used. This gives a wave height of 4 feet and a period of 5 seconds for the significant wave.

A somewhat simpler method has been devised for coastal

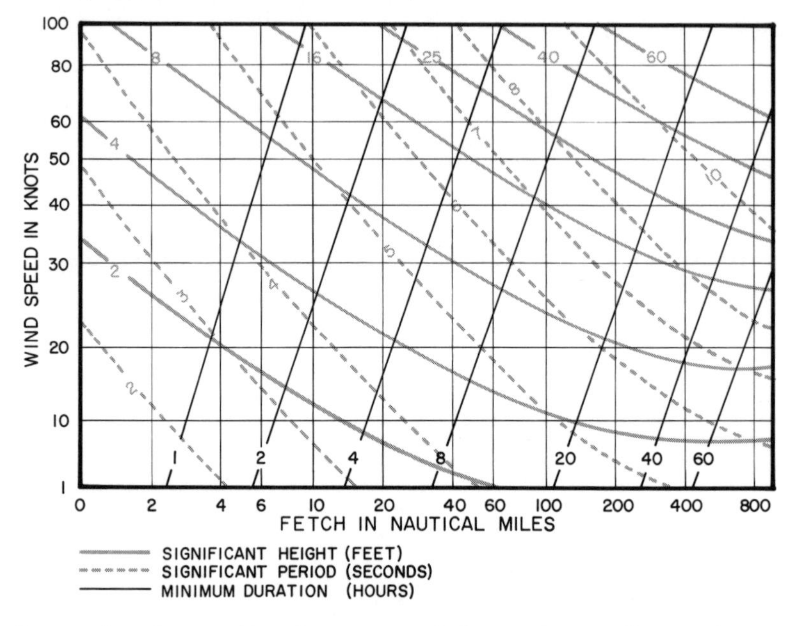

Forecasting the sea: Based upon actual observations a diagram is constructed that enables one to forecast the probable sea to be expected under various conditions. In order to use this, the wind speed is found on the vertical scale to the left. This is then traced horizontally across the diagram until it intersects with either the fetch (read from the horizontal scale at a point vertically beneath) or the duration (on the oblique lines), whichever is met first. At the point of intersection the forecast significant wave height and period are read from the nearest heavy curves. For instance, with a fetch of 40 nautical miles, as in the Straits of Florida between Miami and Gun Cay, and a wind of 20 knots, the point of intersection is between duration lines of 2 and 4 hours, and to the blue lines for 5-second waves and between the lines for 4-feet and 8-feet significant height. So long as the wind has blown for 3 hours the forecast will hold. The waves will not be greater for longer duration, but for shorter duration the intersection of wind speed and duration must be sought. (After Bretschneider, B.E.D.)

waters, in which waves do not behave quite the same as in the open ocean. This is based upon the highest wave observed during a 10-minute period. It is higher than the significant height by a factor of 1.2 to 1.5. It is more meaningful to the average seaman than the highest of the highest third, or significant wave, and easier to observe. Using the simpler diagrams produced for British coastal waters, the mariner uses the first diagram for wave height. Here he finds that the intersection of a 20-knot wind speed and a 40-mile fetch coincides with 6-hour duration and maximum wave of 6 feet in 10 minutes. In the companion diagram the wave period is similarly found to be between 5 and 6 seconds. This agrees fairly closely with the more complicated chart previously referred to. It will be noticed that the forecast highest wave

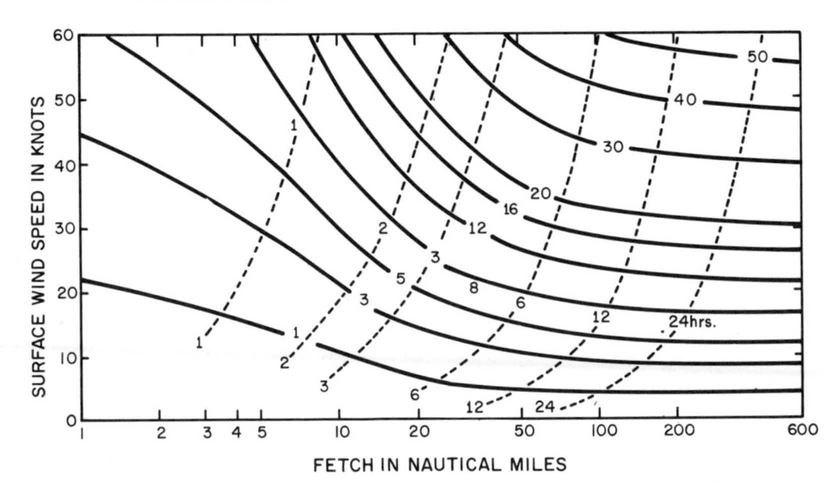

Maximum heights of coastal waves for any given wind strength, fetch, and duration. To use the diagram, the estimated surface wind speed in knots is found on the vertical scale to the left. This is then traced horizontally across the diagram until it is vertically above the known fetch in nautical miles, shown on the horizontal scale. The figure for maximum wave height in feet to be expected in a 10-minute period is read off the nearest whole line curve. If between two curves, a height between the two figures is used. If, however, the horizontal line for wind speed intersects with the broken curved line for the known wind duration in hours, before reaching the fetch, then this intersection point should be used. Whichever intersection point is reached first, fetch or duration, is the one used. (After L. Draper)

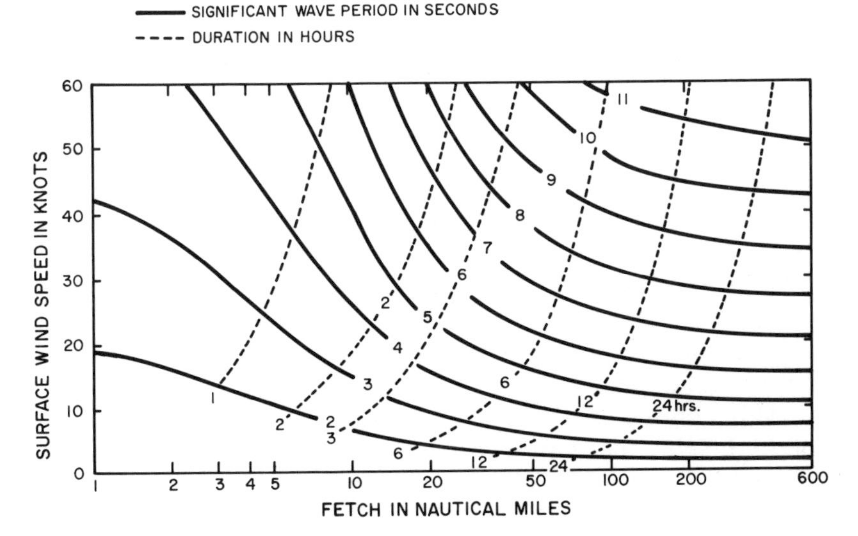

Significant wave period in seconds for given fetch or duration. This is used in exactly the same way as the previous figure for wave height. The whole line curves are for wave period, the broken ones are for wind duration. (After L. Draper)

height for a 10-minute period is higher than the significant wave height, as determined in the previous example.

In the Straits of Florida, it should be added, the foregoing forecast is good only for easterly or westerly winds blowing at right angles to the Straits for which the fetch is 40 miles. With westerly winds the forecast would be for the sea developing at Gun Cay, for easterly winds the sea near Miami. At other places the seas would diminish as one progresses to windward. With winds in the north or south, however, the fetch would be much greater and duration would probably be limiting. Also, a northerly wind would blow against the stream and develop a steeper sea than forecast. With southerly winds the result would be a lower sea.

There is always a statistical chance for a much higher wave than the average. When crests of a number of waves of different period happen to coincide they combine to form a higher wave. The chances of this happening, for any particular height, can be calculated. The chances of a wave twice the average height occurring is about one in 25 but the chances of a wave four times the average is one in more than 250,000. Such high waves are very unstable, so that their crests are quickly blown off and they therefore have a very short life. Although the chance of meeting such monsters is very slight, it nonetheless exists.

The longer a vessel remains in a storm area, the greater is the probability of a larger wave being experienced. The possibility of this happening may be found in the third diagram, which gives a factor for multiplying the wave height (the highest wave in 10 minutes obtained from the forecasted wind conditions) to obtain the maximum height to be expected during a given number of hours of exposure to the storm.

Although the energy in a wave is proportional to the square of its height, the danger from exceptional seas depends not so much upon size as upon steepness. A wave 300 feet in length traveling at more than 20 knots will not damage a ship so long as it is not breaking or near the point of breaking. The ship simply rises to the sea. If the wave breaks, however, the water travels at the crest at greater than orbital velocity and

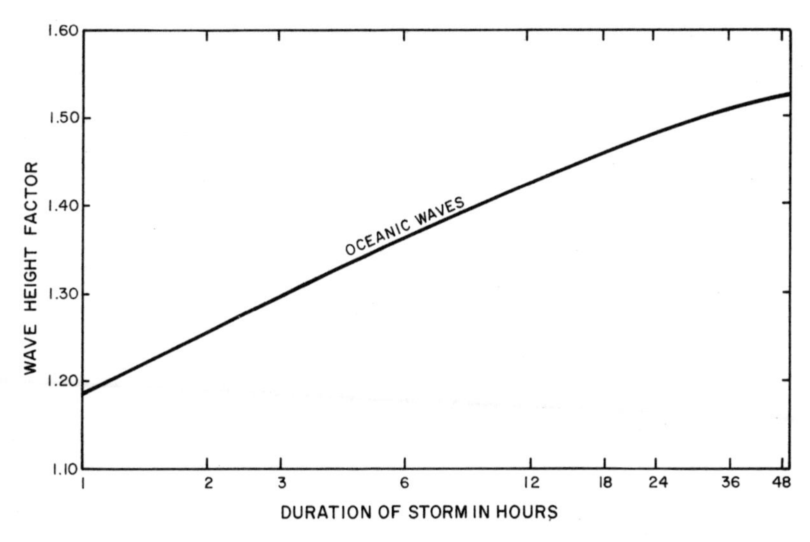

To predict the maximum wave height that might be expected at some time during the entire period of a storm, a point is found on the oblique line immediately above a figure for the duration of the storm in hours (bottom scale). Looking horizontally from this point to the left-hand scale will be found a figure for the wave height factor. This figure is used to multiply the maximum 10-minute height found in the diagram at the bottom of page 49.

the resulting collision with the ship could involve a considerable amount of energy, equal to that of a solid body of the same weight as the water traveling at more than 20 knots. That serious wave damage occurs to modern ships is exemplified by the loss of three destroyers in a cyclone in the Philippines in 1944 and by the accident that befell the U.S. heavy cruiser *Pittsburgh* in 1965. During a typhoon, a heavy sea ripped off 90 feet of her bow. In April 1966 the Italian passenger liner *Michelangelo* was damaged by a wave that smashed windows 81 feet above the waterline, flooded the forward half of the ship, tore off the heavy metal flaring on the bow, and killed three people.

Waves higher than 45 feet are not often encountered in the North Atlantic Ocean, yet a number of exceptional cases have been reliably reported, along with others that are probably erroneous. Estimates of mountainous seas are often exaggerated by mariners, quite unintentionally, because of the difficulty of making measurements of wave height without the aid of suitable instruments. Without such equipment the simplest and most accurate method of measurement is for the observer to mount to successive higher parts of the ship, such as the main deck, the bridge, a mast, the rigging, or the crow's nest. At an appropriate height, when the ship is in a trough and on an even keel and the crest of the nearest wave is just beginning to obscure the horizon, the wave height is equal to the height above the waterline of the eye of the observer.

One of the reasons that wave heights are apt to be overestimated is that broken water from a wave of moderate height may be swept up as high as the bridge, far above the actual wave height. When green, or solid, water is taken over the bow or sides of a ship as she plunges, the wave height appears to be much greater than it actually is.

A sea with 60-foot-high crests, caused by waves crossing from different directions, was reliably reported in 1848. Another wave that was measured with reasonable accuracy by competent observers was the monster encountered by the liner *Ascanius* while on a passage between Yokohama and Seattle in 1921. During an extended period of hurricane winds, while the ship was in the trough and on an even keel, ship personnel at an eye level 60 feet above the water surface noted that the wave crests obscured the horizon. The waves could have been as much as 70 feet high. Similarly in 1923 the liner *Majestic* experienced winds of hurricane strength in the Atlantic. The seas were very regular but of phenomenal size. On this occasion the wave crests obscured the horizon for an observer whose eye was 89 feet above water level. However, the ship was pitching at the time and, in the opinion of an expert, a fair estimate of wave height might be about 75 feet.

The highest recorded wave so far was one of a series observed from the U.S.S. *Ramapo,* a 478-foot naval tanker, while en route between Manila and San Diego during February 1933. A wind of about 60 knots was blowing from astern, and huge seas had been running. The seas were not breaking; there was no rolling and easy pitching. The longest period was nearly 15 seconds, from which the wavelength may be calculated as about 1,100 feet and the wave speed as more than 50 knots. The watch officer later noticed an increase in wave heights in a succession of waves at 80, 90, 100, and 107 feet, and finally the enormous height of 112 feet. The last estimate, though at first unbelievable, appears to be fairly reasonable, since the ship was not listed, her stern was in the trough, and the officer saw the crests astern in line with the crow's nest and the horizon at the time of observation. Knowing the length of a line from eye to crow's nest and the angle made with the ship's horizontal, it was a matter of simple trigonometry to calculate the wave height.

Obviously a wave of such dimensions could not result from a storm with the usual duration and fetch. It turned out that the storm was unusually widespread, with an unobstructed fetch of thousands of miles and a duration of seven days. Possibly also a combination of wave trains of different periods helped to bring about such unusual seas. If this wave appears phenomenal, it must be added that the longest recorded swell, though not as high as the foregoing, was reported as being 2,600 feet long from crest to crest with a speed of more than 70 knots.

While there is considerable interest in freak waves of exceptional height, even sailors are not always cognizant of the possibilities of exceptionally deep troughs. Two interesting cases of this have been reported, both occurring off the coast of Africa between Durban and Cape Town. In August 1964 the *Edinburgh Castle*, 750 feet long and of 28,600 gross tons, was steaming into a heavy southwest swell without any difficulty. The master changed course, however, to take the swell off the bow instead of driving into it head on, with the result that the movement became more comfortable. The wave-

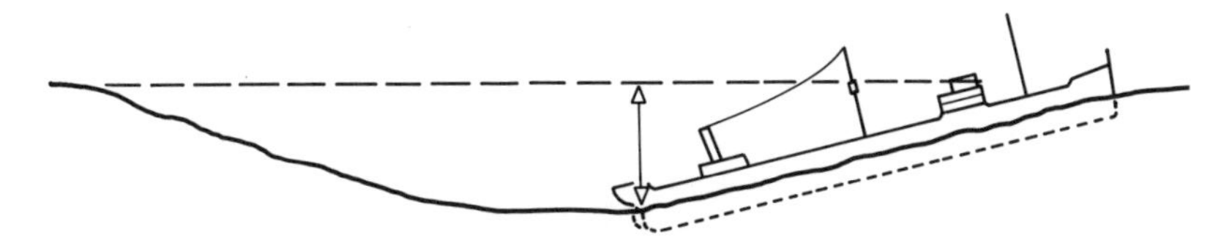

The sketch shows the manner in which the giant wave was measured from the bridge of the U.S.S. Ramapo. The observer found the crow's nest in line with the wave crest behind the ship, and also with the horizon. Since the stern was in the trough and the heights of the observer's eye and of the crow's nest above the waterline were known, the wave height could be calculated.

lengths were about 150 feet and the ship was pitching from 10 to 15 degrees from the horizontal. Suddenly, as the ship lifted, the next wave appeared to be about 300 feet long, and she charged at an angle of 30 degrees down into a hole in the ocean. Before she could lift again, the next wave was shoveled aboard as solid water to a height of 15 to 20 feet. Athwart ship rails and a ladder were carried away. In the same general area, the British cruiser *Birmingham* also hit a hole during World War II, while steaming comfortably in a moderate sea and swell. When she hit the hole, suddenly and without warning, the next sea came green over A and B turrets and broke over the open bridge, 60 feet above the waterline. The pitch darkness in which she was steaming and her blacked-out condition added to the confusion. In both of these cases it is possible that the hole was caused by the coincidence of the troughs of a number of different wave trains, just as in the case of high crests.

To yachtsmen cruising the open oceans, and more especially during ocean races when men and gear are closer to the limits of endurance, the stress of the normally high waves in a gale are naturally considered, but the possibilities of a sudden destructive freak sea cannot be neglected.

In addition to the question of the safety of ships at sea, ocean waves are of concern to man because of the need to develop the best procedures to follow on the fortunately rather rare occasions when disabled aircraft must be ditched.

The advice usually given for ditching seaplanes is that unless the wind speeds are prohibitive the craft should attempt to land crosswind immediately behind the crest of the major waves or swells or into the trough; in both cases taking care to use the controls to compensate for weathercocking effect and the downwind travel of the wave crests. The next best procedure, if crosswind landing is impracticable, is to land downwind just beyond the crest of a wave. Landing into the wind is the most hazardous, since the oncoming waves have the effect of rapidly repeated blows upon the airframe. The problems of takeoff are rather similar, with takeoff parallel to the wave crests recommended. In all cases it is advised that the pilot seek the patch of lesser seas that often alternates with groups of higher seas. These recommendations were developed in the days of propeller-driven seaplanes. Today, jet-engined land planes with high landing speeds have their own special problems, but the record of ditchings with no loss of life is still high.

The biological oceanographer is also interested in the effect of waves upon life in the sea. An example of this interest has to do with the effects of orbital movements of water and the breaking of crests, which together bring about a mixing of the surface and lower layers of water to a depth depending upon wavelength, in which heat, oxygen, and nutritive elements are evenly distributed. In spring, generally, winds are lighter and seas smoother. The solar heat entering the ocean surface is distributed evenly to a comparatively shallow depth as a result of the mixing effect of waves. Below this depth there is a rapid drop of temperature with depth known as the thermocline. Above the thermocline, microscopic plant life, energized by strong solar light, blossoms at a greater pace

A knowledge of the mechanism of wave action is helpful to the pilot when he is forced to ditch a plane at sea. The proportion of ditchings that have been carried out successfully without loss of life has been high. (U.S. Coast Guard)

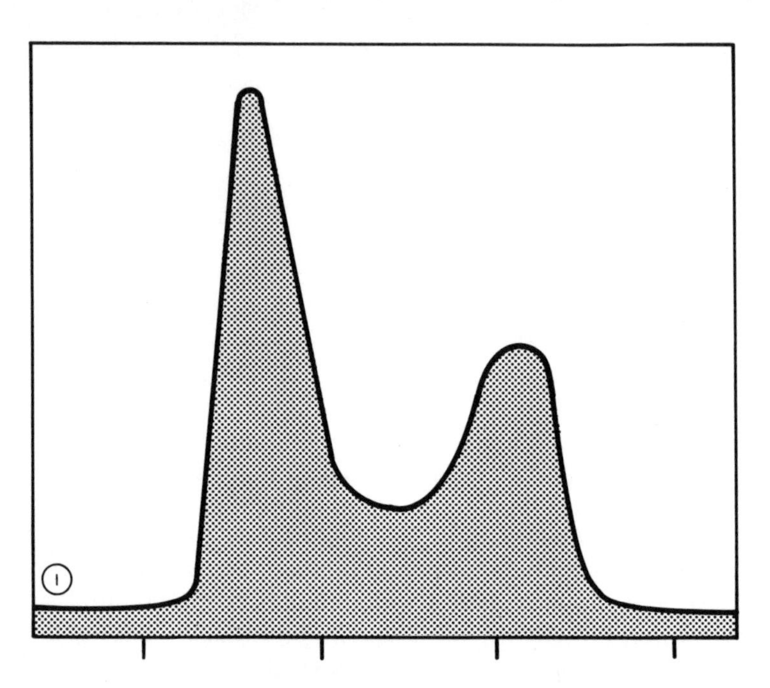

Relationship between quantities of phytoplankton; nutrient salts in solution, during seasonal changes in light intensity, water temperatures, and turbulent mixing. The upper figure shows amount of phytoplankton; the lower figure, light intensity (dotted line) and nutrient salts (shaded area). For further explanation, see text. (After Russell & Yonge)

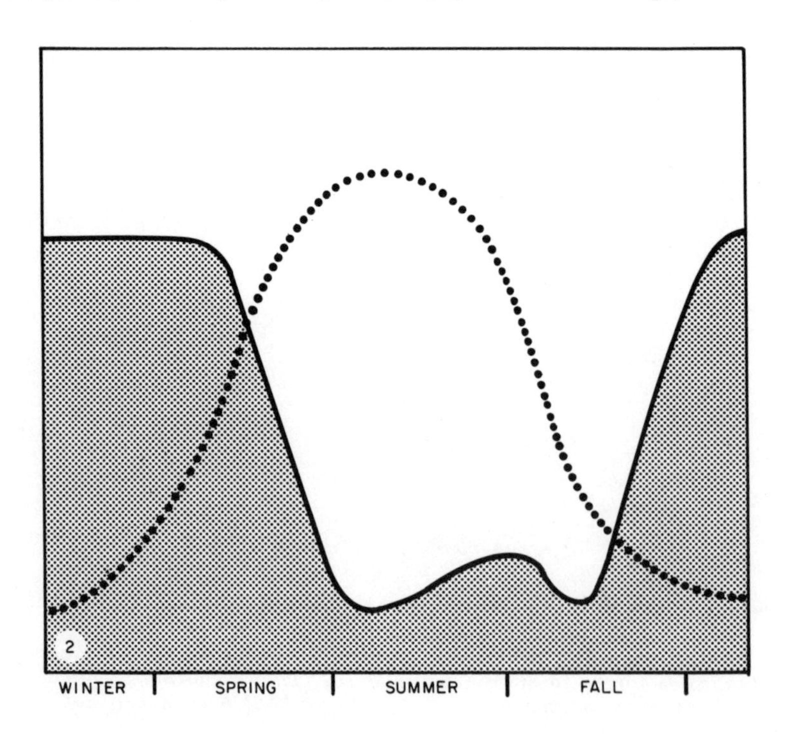

in the spring, until it uses up the dissolved fertilizer salts, phosphates, and nitrates. The death of the plants and the animals that feed upon them releases organic materials into the water as they sink below the thermocline. Bacterial decay converts these once more into fertilizer materials, but only the portion released above the thermocline is available for continued growth. Accordingly, lacking nutrient sources, during the summer the great spring flowering begins to dwindle.

In the fall and winter, heavier seas and the deeper influence of wave action break up the thermocline and restore nutrients to the upper waters. But by this time the water is cooler and the sunlight weaker, so that plant life continues at a much lower level of production. The cycle is renewed in the spring, when sunlight increases and the temperature rises, with the nutrients still remaining in the upper layers. This cycle is one of the reasons that much of the open ocean is very unproductive of life. The summer thermocline permits the downward escape of organic detritus to a depth below the thermocline from which wave action cannot retrieve it. Only in shallow seas where the wave action of heavy winter seas extends to the sea floor are all of the nutrients restored to the surface. There are places in deeper water, however, where other factors, such as vertically moving currents, are effective in bringing the deeper nutrient-laden waters to the surface, where sunlight is available for the photosynthesis upon which plants depend for growth. These places are usually areas of high productivity.

5. *Shallow-Water Waves*

As mentioned in Chapter 3, as seas travel outward from a storm area toward the land and shallow coastal waters they begin to lose energy. The shorter waves decay most rapidly, but the longer ones travel as swell, losing height but gaining length as they go. The loss of energy is partly the result of turbulent friction. The effects of this on small waves may be observed in the turbulent area in the wake of a ship. Apart from the stern wave, the general surface of the sea is markedly less choppy than in the area outside. Also, when a ship is drifting with the seas abeam, the area immediately to windward is less choppy than the general surface, probably because of the turbulent displacement of water as the ship drifts downwind. A major factor in energy loss is the divergence of the wave front. The waves radiate over a wide sector. As the front lengthens the energy becomes distributed over a greater area so that the energy in the unit area decreases. If there are no winds, about one-third of the wave height is lost each time the wave has traveled a distance in miles equal to the wavelength in feet. Thus, a 1,000-foot wave 30 feet high will become 2½ feet high after traveling 6,000 miles, unless it enters shallow water.

Before the swell dies or reaches shallow water and breaks, it may travel enormous distances at the speed of the wave train, which is half that of the individual crest. For instance, a swell breaking on the island of St. Helena has been traced back to a North Atlantic storm more than 4,000 miles away. Other examples include swells that have traveled 4,000 to 7,000 miles through the Tasman Sea and the Pacific Ocean to California. The appearance of a swell, therefore, may indicate a distant storm. If, however, the swell begins to increase in wave height it may be a warning to the mariner that the storm in which it originated is approaching him.

When the swell approaches a distant coast, its measurement

and analysis may be used to locate the distant storm. Even if this should be in a relatively untraveled area of the ocean, the information could be most valuable to ships outside the storm area, since the storm might be moving toward more heavily traveled shipping lanes.

When evidences of a distant storm arrive at a coastline, they may be recorded on wave recorders and analyzed to give a spectrum. The first evidences of the storm are the longest and lowest waves. The diagram shows swell arriving at the coast of Cornwall, England, on March 14 and 15, 1945. First, at 1 P.M. on the 14th a low swell of 24-second duration appeared. Gradually, shorter waves of greater height arrived, until at 11 A.M. on the 15th, waves of 10-second period made their appearance. Since the shorter swells arrive later than the longer ones, because of their known group speeds the difference in time of arrival may be used to estimate the distance of the storm by a simple calculation.* Since swell tends to increase in period and decrease in height the longer it travels, the ratio of height to period also gives an indication of the distance of the storm. Measurements at different places enable not only the distance but also the direction of the storm to be determined by triangulation. It was found by calculation that the swell mentioned above originated in a storm to the southeast of New Foundland about 2,000 miles away. In another case, a swell arriving off Cornwall on June 30, 1945, which had a period of 18 seconds, was traced to an intense tropical hurricane between Cape Hatteras, North Carolina, and Nantucket, Massachusetts, about 2,500 to 3,000 miles away.

As the swell approaches shallow water, several changes take place. As explained earlier, changes occur in the speed, length, height, and shape of waves, but not the period. There

---

* Method of estimating location of storm origin of swell by difference in time of arrival of long and short swell.

Long swells of length L_1^{feet} traveling at C_1 knots.

Short swells of length L_2^{feet} traveling at C_2 knots.

Difference in time of arrival of swells, t hours.

Distance of origin $= t \left(\frac{C_1C_2}{C_1 - C_2}\right)$ nautical miles.

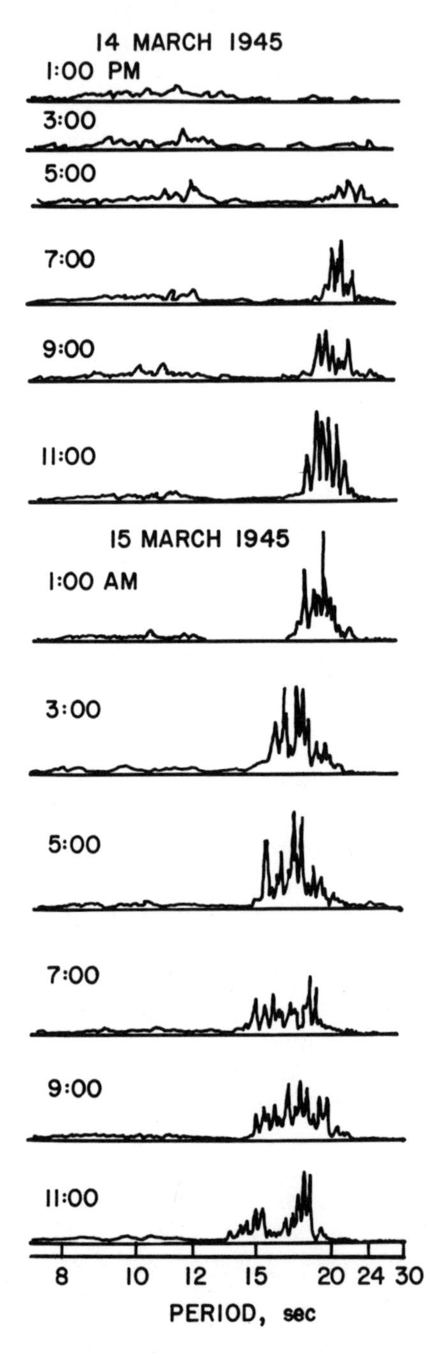

Evidence of a distant storm: Wave spectra from records taken of swell arriving at the coast of Cornwall between 1 P.M. *on March 14, 1945, and 11* A.M. *the next day. It will be noted that swells of low height, but periods of 24 seconds, began to arrive at the beginning of the period. The energy increased, but the period gradually decreased until the maximum was concentrated in 18-second waves with a secondary maximum in 15-second waves. (After Deacon)*

is also a change in direction. Finally, when the water becomes sufficiently shallow, the wave breaks.

Waves entering shallow water slow down because of the narrower vertical space available for orbital movements between the surface and the sea floor. With less room available, the orbits become more and more flattened. The speed no longer depends upon the wavelength, but only upon the depth. Consequently, as the depth of water decreases, the velocity of the wave also decreases.

When a long wave crest enters gradually shallowing water at an angle to the shore, the end nearer shore will enter the shallow water first and will be slowed down. The rest of the wave crest, still in deep water, retains its original speed until it, in turn, reaches shallower water and slows down.

This situation results in a bending, or refraction, of the wave front, similar to the refraction of light in passing from air through a glass lens. The refraction of waves accounts for the fact that on an evenly sloping shore the waves always tend to arrive more or less parallel to the beach, no matter how great the angle at which they originally advanced from the deep ocean. It also explains the movement of waves around a headland into a bay. In many places a bay or natural harbor anchorage that is sheltered from local storm waves of comparatively short length will nevertheless be subject to the action of swell. This is because the longer waves are slowed down by shallow water to a greater extent than short waves. The long waves are bent by refraction inside the harbor entrance and so spread out behind the breakwater, while the shorter waves continue in a straight path inward from the entrance.

An important result of refraction is the redistribution of wave energy, which may become concentrated against headlands but diffused in bays. The redistribution of energy explains the frequent occurrence of heavy wave action in the vicinity of headlands, whereas the intervening shores of bays may be comparatively calm and provide safe anchorage. It may be demonstrated by drawing orthogonals, or lines that run at right angles to the wave crests. If it is assumed that waves approaching from a distance have straight, uniform wave crests, lines drawn at right angles to the crest, at equal distances apart, will enclose areas of equal energy. So long as the wave crests are parallel to the shore and the bottom slope is uniform, the isogonals run in straight lines perpendicular to the shore. If, however, the bottom slope is not uniform but includes a submarine spur running out to sea from the headland, the wave front in this area is slowed down and becomes curved. The orthogonals, still drawn at equal distances apart, now converge toward the headland. Although the energy between the orthogonals does not change to a great extent, it is now crowded into a smaller area and, therefore, results in higher and more destructive waves. The reverse is true in the bays, where the orthogonals diverge and the energy is diffused over a greater area. The closeness of the orthogonals is a measure of the energy concentration.

In places where ravine-like submarine canyons occur offshore, the whereabouts of the canyon is disclosed to an onlooker ashore by the relative absence of high waves in the area. In this case the waves slow down on either side of the canyon but advance at normal speed above it. As the resulting wave front bulges toward the shore over the canyon, it lengthens and loses energy. If the orthogonals are drawn perpendicular to the wave crests they will be found to crowd together outside of the canyon area but diverge over the canyon itself. Similarly, an offshore shoal, running out from the land, will reveal itself by increased wave action, as the waves slow down, converge, and steepen in the shallow water.

Yachtsmen naturally prefer to anchor in the lee of an island if there is much wind and sea. If the sea consists predominantly of short period waves, conditions in the lee may well be calm. If, however, the sea has had time to develop longer waves or there is a big swell, the anchorage may be very uncomfortable, though safe, because of the refraction of the long period waves that come into the shadow of the lee. Refraction effects such as this may result in changes in wave direction for a great distance beyond the island. The resulting cross patterns are believed to be the means whereby the Polynesians performed their amazing feats of navigation across the Pacific Ocean. Their "maps" consisted of sticks

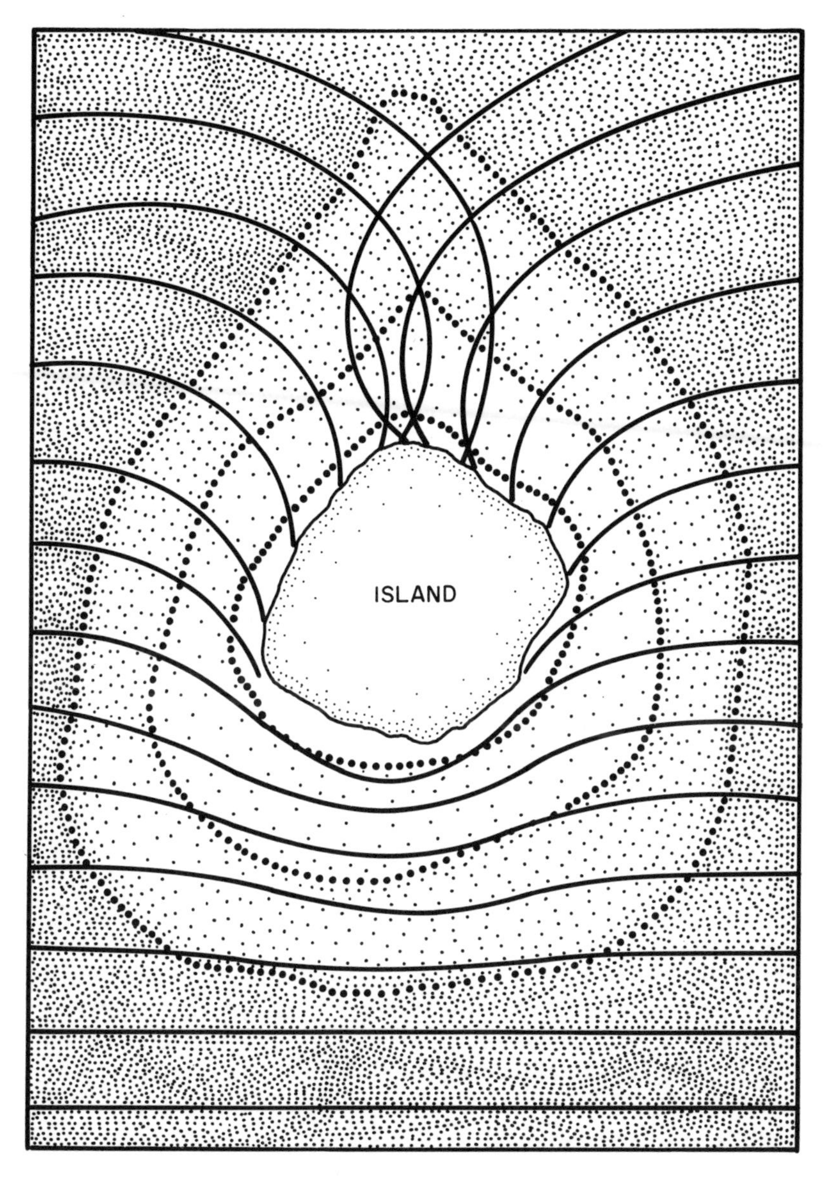

Refraction of waves around an island. As a wave approaches, the part first entering the shallower water is slowed down, so that the wave front curves around the island. This continues until two wave crests meet each other in the lee. Knowledge of this effect, which may be found even well out of sight of the island, was used by Polynesian navigators to find their way across the Pacific Ocean without instruments.

tied together to show the pattern of waves interfering with each other as a result of wave action around distant islands.

Since the probable effect of wave refraction is of considerable practical importance when planning breakwaters or similar installations, a wave refraction diagram is a useful tool.

Plotted on a chart of suitable scale, with bottom contours marked at intervals, the wave refraction diagram may be constructed for the arriving wavelength most likely to predominate. Sometimes diagrams are constructed for a number of different directions and wavelengths, representing the various conditions that might be encountered.

A wave diagram may be prepared with simple navigational drawing instruments such as a compass and a rule. First, where the chart shows a depth a little more than one-half of the deepwater wavelength, a straight line may be drawn perpendicular to the assumed direction of the wave. This is marked with a series of points at equal distances apart. Calculations are then made for the wavelength of each point, based upon the actual depth. As explained in Chapter 2, the wavelength in shallow water is related to the depth. The new crest is then drawn as a tangent to arcs struck at each of the points on the crest, using the calculated wavelengths as radii. At each of the new points, where the tangents touch the arcs, a similar procedure is followed, using the new wavelengths calculated for the new depths. When this has been repeated over the entire path of the wave, the orthogonals are drawn in by eye, perpendicular to each wave crest.

This process explains the manner in which aerial photography of waves in shallow water enabled the planners of beach invasions during World War II to estimate the bottom contours and depths of water in the approaches to the shore. First, the wavelength may be measured from a suitable scale, based upon the height of the airplane. From positions of a crest in successive frames of the film, the velocity of the wave may be calculated. The dependence of velocity upon depth enables decreasing depth to be estimated from the decreasing wavelength or velocity. In addition to this, the crowding of waves over a bar or the spreading out of waves over a deeper area indicate the presence of these irregularities.

Another mechanism whereby a wave front may change direction and spread behind an obstacle, such as a breakwater, is diffraction, which is analogous to the diffraction of light around the edge of an object or through the slit of a spectrometer. Changes in wave direction of this type are inde-

Wave diffraction demonstrated in ripple tank. Waves advancing from above reach an opening, simulating the entrance to a harbor. When the wavelength is long compared to the size of the entrance, as in the case of swell, diffraction causes the entering wave to spread out over the harbor in a circle (left). When the wavelength is small, compared to the entrance (right), as in the case of local storm waves, the waves are less diffracted and tend to advance in a straight front over a restricted area.

pendent of depth and take place even where the depth of water is greater than a half wavelength. At some distance sideways from the end of a breakwater, for instance, an approaching wave will continue on its original path. Nearer the end of the breakwater, however, the wave will begin to curve into the lee of the structure. Since the wave crest is thereby stretched, the energy of the curving wave is necessarily decreased. Diffraction explains the action of longer waves that continue to advance even in deep water harbors into the shelter of the obstruction, though with diminished intensity. The amount of diffraction, or bending, increases as the ratio of the opening in the breakwater to the wavelength decreases.

Diffraction of waves by an obstacle is greater for waves of larger wavelength than for waves of shorter length so that, just as in wave refraction, the obstacle, such as a harbor jetty, may provide better shelter from the shorter seas than from the longer swell.

Water waves are not only refracted and diffracted in the manner of light waves, they are also reflected in a similar fashion. A wave advancing in water of fairly uniform depth will reach a seawall or other impervious vertical obstacle and be reflected. If the wave advances at an angle, the reflected wave will pass through the advancing incident wave with little change, forming a moving pattern of diamonds, lozenges, or squares, depending upon the angle. This can be readily demonstrated in a swimming pool. Even on a beach with a fairly steep slope it may sometimes be noticed that a single wave or a train of two or three waves returns to the sea from the beach at an angle to the incident waves. If the wave advances against the reflecting surface head on, then a standing wave, or clapotis, may be formed. This happens when the opposing wave crests and lower troughs alternately coincide and oppose each other, causing higher crests and lower troughs than in the original wave. The crests and troughs of the clapotis do not move horizontally, but at places one-half wavelength apart, the sea alternately rises upward as a high stationary crest and falls as a low trough. Standing waves of this nature, but on a larger scale, are important features of some tidal systems and are described in Chapter 11.

Wave reflection in swimming pool. A wave was generated at an oblique angle to the side nearest the reader and was reflected from the far wall. The two waves interact to form lozenge-shaped patterns and show as bright lines on the floor of the pool.

Although deepwater waves entering shallow water gain ultimately in height, a careful observer will notice that on a uniformly sloping bottom they first diminish slightly in height, at a point where the depth becomes less than half the wavelength. This results from the fact that whereas in deep water the energy, and therefore the group, travels at half the speed of the individual wave, in shallow water the energy now begins to move at the wave speed itself. Not only is the potential energy carried with the wave, but part of the kinetic energy is transferred also. The wave begins to take on a little of the character of a solitary wave, in which the crests become more and more separated from each other and are virtually independent of the trough, each crest moving as a ridge of water across an otherwise more or less level surface of water.

After the initial reduction of height as a result of the increased transfer of energy, the wave begins to increase in height as it enters a water depth of less than $1/20$ of the wavelength. This happens because of the reduced wavelength and consequent crowding of energy into a smaller area. The waves also become more asymmetrical, that is, the front slope of the wave becomes markedly steeper than the rear slope. Although this increased asymmetry has been described as a result of bottom friction, it may be caused by the fact that the crest is in deeper water and therefore travels faster than the preceding trough, which it begins to overtake. It has also been pointed out that the forward movement of particles in the crest is in the same direction as the wave movement and so continues longer than the backward movement in the trough, which is opposed to the wave movement and is therefore of shorter duration. In addition, the kinetic energy of particle movement in the crest may be slowed less rapidly than in the trough, where the effect of decreasing depth is sooner felt.

As a wave continues to advance into shallower water its height increases until it exceeds about two-thirds of the water depth. The exact ratio varies with the type of bottom, its slope, and the initial steepness of the wave. At this point the forward orbital velocity of water particles at the crest increases beyond the velocity of the wave and the wave breaks.

Spilling breakers, in which the wave crest breaks up before overtaking the forward slope. The broken water runs down the slope.

Plunging breakers, in which the wave steepens and the crest arches forward as a solid arch of water, which often forms a tunnel of air beneath it. (U.S. Coast Guard)

The manner of breaking varies also, from the massive plunging type, in which the crest arches forward over the trough and falls with a roar, imprisoning a pocket of air below it, to the less dramatic spilling type, in which the crest collapses more quietly into a mass of foam rolling forward toward the beach. Between the two extremes there are forms of breaker intermediate between plunging and spilling. In fact, it may frequently be noticed that the same wave plunges along part of its length but spills in other parts. The type of breaker is determined partly by the slope of the sea floor and partly by the initial form of the wave. A high symmetrical crest traveling over a gently sloping bed will gradually reach the critical height, where it will begin to spill more or less gently forward. A lower crest moving in over a steeper bed will become less symmetrical, with the forward face becoming steeper until the crest plunges violently forward. The wave may break prematurely if strong offshore winds blow the crest forward or if there are irregularities on the bottom.

In some cases, where there is an offshore bar, waves may break, only to re-form again at some distance closer to shore, as the wave again enters a shelving bottom. This is more likely to happen with spilling than with plunging breakers, since the latter lose the greater part of their energy the moment they break, whereas the former may retain sufficient energy to again increase their height to breaking point when the depth becomes shallow enough.

When a wave breaks on a beach with very gradual slopes, it may be noticed that a small wave emerges and travels rapidly toward shore, independently of other waves and separated from its successors by a long stretch of level water. This is a true solitary wave, in which movement occurs not in the trough but in a forward movement of the crest alone. Such waves are sometimes called waves of translation, since the water in them is moving with the wave itself rather than in stationary orbits.

Along the shore groups of higher waves frequently alternate with groups of lower ones, giving rise to the idea that a fixed number of waves intervene between every high one. As in the case of waves at sea, the number is not a fixed and invariable one. Nevertheless, the periodical increase in wave height is a common occurrence. It may be caused by the interference of waves of different periods. It may also result from interference from low waves of much greater length than the majority. Another factor that may enter into this situation is the periodical movement of water that has accumulated near the beach as a result of the breaking of waves.

As the water returns to the sea the adverse current may slightly increase the tendency of waves to increase in height and breadth. These groups of higher and lower waves, called surf beats, are of importance to fishing boats or lifeboats that must land on or set out from beaches through the surf. Not only is the area sought where the bottom contours cause spilling rather than plunging breakers, but the period between wave beats is timed in order to move through the period of lower waves.

A phenomenon associated with waves breaking on a beach and frequently misunderstood is that of the rip current. When a bather, standing on the bottom in front of an advancing wave about to break, feels his legs pulled out from beneath him by an outgoing surge of water, he is under the control of the strong backward movement of the trough. He may then be bowled over by the oncoming wave and suffer temporary discomfort but he has not been exposed to a rip current. The rip current is more dangerous, but if the bather understands it and is a reasonably good swimmer he need not be in great danger. Typically, the victim finds himself drifting out to sea and attempts to swim directly back to shore, fighting a current setting offshore, against which he is unable to make headway. Under these conditions he is liable to panic, redouble his efforts, and weaken. Rather than attempt a direct course to the shore, the swimmer should swim parallel to the beach until he is clear of the outgoing current and then with the assistance of the incoming waves find his way to shore, as will be understood from the following.

The cause of rip currents lies in the fact that breaking waves produce a net transport of water beachward. Measurements show that at various intervals the average sea level some distance offshore may become periodically lower than that nearer the beach because of the landward transport of water. The phenomenon of surf beats illustrates this rhythmic carriage of water to the beach, followed by a compensating return flow to the sea. After the arrival of a group of waves of greater than usual height, it will be noticed that the wetted surface of the beach is higher than it was before or after. Since the wetted level recedes between beats, the water must somehow have escaped back to the sea. Any bather who takes the trouble to throw a beach ball or other floating object out into the breaking surf will find that it will often move in a direction parallel to the beach for a distance and then move outward toward the ocean. If the waves do not approach the beach completely parallel to it, the longshore wave drift will travel only in one direction parallel to the beach before moving out into the rip current. Where the waves closely parallel the shore, the drift may move toward the rip currents from both directions. Since the breaking waves stir up the sand, the outward moving rip currents may be clearly marked by the muddy appearance of the outgoing water stream.

As the water piles up on the beach it will naturally move to left or right until it finds an easy path out to sea again. This is usually where a local increase in depth occurs, such as a break in an offshore bar. There the waves are less liable to break, so that transport beachward is at a minimum and the accumulation of water on the beach is able to find its way back to sea. This is the course the bather who is having difficulty should seek. Since the rip current flows offshore at a point where the waves are breaking less, the rougher water is more likely to be free of offshore currents. If help is at hand, it is better for the swimmer to remain beyond the breaker zone altogether, where the rip currents die out.

The surf rider takes advantage of the steep-crested waves and a helpful balance of the various forces involved. Although the plunging wave is more spectacular, the longest ride is on a spilling wave, or one intermediate between spilling and plunging. The rider, just in front of the wave crest, is at the steepest part of the wave slope. Gravity helps him slide down the hill ahead of the breaking crest. Also helping is the orbital velocity of the water particles, which, near the crest, are moving forward. If the surfer slides down too rapidly toward the trough, he advances into a lesser slope, the gravity pull and the orbital velocity become less in the direction of travel, and, consequently, the wave begins to overtake him. As the crest begins to overtake him, the combined action of slope and water movement again tend to carry him ahead of the wave. The prime difficulty is that with the water moving at almost wave velocity, he is moving at nearly the same speed as the water and therefore has little steerage way. By moving somewhat laterally, instead of directly before the wave, he is able to gain some steerage way through the water, and the fin below the surfboard will also help to keep him on a steady course. This lateral movement has another advantage. Al-

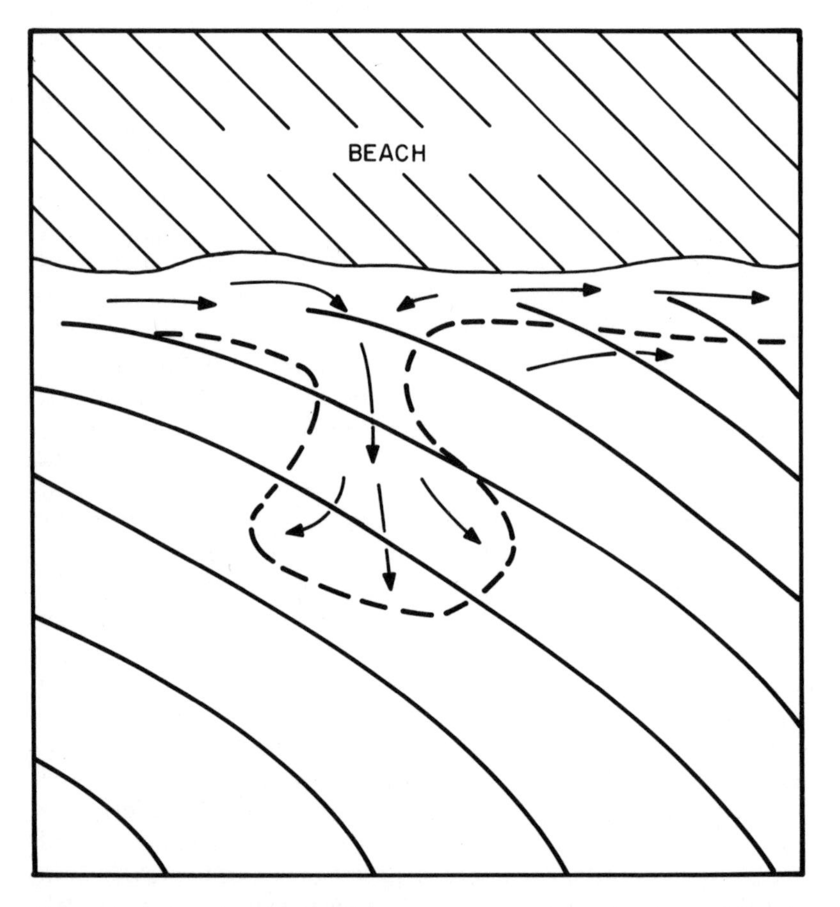

Formation of a rip current. The waves approaching the shore obliquely generate a current or drift along the beach. At a point along the shore where the bottom deepens, or there is a break in an offshore bar, the water finds its way back to sea as a rip current. If caught in such a current, the swimmer should swim parallel to the beach until beyond the rip, when he can then swim ashore.

though waves tend to curve in shallow water until they become almost parallel with the beach, they do not often become completely aligned. Often one end of a long crest will reach water of critical depth first and will begin to break. After this, the breaking water will move from one end of the crest to the other, as successive parts of the crest reach shallower water. By moving partially sideways along the wave front the surfer keeps ahead of the breaker. In so doing he enjoys a longer ride, and if skillful, he avoids being engulfed by a plunging crest.

6. *Waves Against the Land*

The destructive energy of wind waves at sea has accounted for the loss of and heavy damage to thousands of ships. The storm waves that break against a rocky coast, however, the result of waves colliding not with a yielding vessel but with a stationary cliff or rigid man-made structure, together with the tremendous concentration of energy, produce even more impressive displays of wave force. This is true even when, in the absence of local storms, swell from a distant source approaches the shore.

Damage caused by onshore waves ranges from complete or partial destruction of solid masses of masonry to either the loss of beaches or the filling in of entire harbors with sand. Onshore waves frequently extend to a far greater height than those waves that deluge ships at sea. Although there is only one reliable record of a wave more than 100 feet in height at sea, there are a number of cases of damage caused to coastal structures, such as lighthouses, at heights very much greater than this. The damage in such cases is not necessarily disastrous, for at extreme heights it is probable that the water is broken rather than solid. Nevertheless rocks of considerable size have been projected by waves to the tops of lighthouses.

At Donnet Head Light, in the north of Scotland, for instance, waves sometimes hurl rocks with sufficient force to break windows 300 feet above sea level. Trinidad Head on the California coast is on top of a rocky headland nearly 200 feet above sea level. Nevertheless, in 1913 the keeper reported that in spite of this height above the water a great sea shot up the face of the cliff and ran over until it seemed as if solid water was level with him near the lantern, and the force of the impact upon the lighthouse was sufficient to stop the lens from revolving. At about the same height above sea level Uist Light experienced a great wave that stove in a door.

One of the coasts most plagued with great waves is that of the U.S. Pacific Northwest. In this area, 40-foot-high waves are by no means unusual. Tillamook Light, which is on a rock several miles at sea just south of the Columbia River, is guarded by steep cliffs extending 90 feet above low water. In every heavy storm rocks are hurled from the base of the cliff to the top of the rock. At the height of one winter gale in 1894 a boulder weighing between 100 and 200 pounds rocketed up above the light sufficiently far that in its fall it gained enough momentum to break a 20-foot hole in the roof of the keeper's house. The light is 140 feet above the sea and the roof of the house about 100 feet. On another occasion a rock estimated to weigh a half ton was propelled across the space in front of the building. In 1902 the keeper was amazed to see water thrown up to 200 feet high and apparently descending in a solid mass upon the roof. Some years later the foghorn, also on top of the rock, began to malfunction as a result of being filled with small wave-thrown rocks. It is not surprising to learn that the glass of the lantern has been broken several times.

The effect of wave action in shallow water is not only experienced at great heights above sea level, it may also extend to comparable, but smaller, depths below the surface. Bearing in mind the rapidly decreasing orbital movements of waves with respect to depth, one obtains some idea of the power of coastal waves from fishermen's reports that one-pound rocks have been swept into lobster pots at a depth of 180 feet in the English Channel. Off the coast of Ireland rocks of more than 100 pounds have been known to move under the influence of waves in water of comparable depth. At Petershead 40-ton blocks have been moved at 36 feet below the surface.

Although the height and the depth to which coastal wave action extends are impressive, the titanic forces of such waves are exerted to their utmost at levels closer to that of the sea, where massive structures have been torn apart, moved considerable distances, or otherwise destroyed. One of the best-known examples is that of the Eddystone Light, which stands upon an isolated rock some 12 miles at sea from Plymouth Sound on the southwest coast of England, where it is exposed to the full force of westerly gales riding in from the Atlantic. It has been destroyed not once but several times, only to be rebuilt again. On the opposite side of the Atlantic, a well-known New England light, Minots Ledge, was repeatedly destroyed, even during the time it was being built, only like the Phoenix to arise again. Finally completed in 1851, it was soon completely lost, carrying with it the crew. The structure built to replace it is almost 100 feet high and has survived more than one hundred years. Yet from time to time it is entirely engulfed in waves.

For every lighthouse that has been completely destroyed there are any number that have suffered severe damage, either during construction or after completion. During the construction of an Irish lighthouse, Dhu Heartach, in 1872, stones weighing 140 tons, carefully interlocked and cemented into place, were torn out from the partially built tower nearly 40 feet above high water and carried away.

Breakwaters, from their very function of protecting harbors and vessels from heavy swells and storm waves, are naturally located in extremely exposed situations. Small wonder, then, that they offer some of the most spectacular examples of wave power. The twice destroyed breakwater at Wick Bay in Scotland gives ample testimony to this. In the middle of the past century a huge trench was constructed of cement and rubble at the exposed outer end of the breakwater. As a foundation, three layers of blocks were placed in the trench. They were not ordinary blocks, for each one weighed in the neighborhood of 100 tons. The foundation supported three additional layers of large stones cemented together and above these was a cap of cement rubble weighing more than 800 tons. The foundation and cap were tied together by means of iron rods 3½ inches in diameter. But even this carefully designed mass could not survive the great storm of 1872. The whole outer end of the breakwater, 45 feet long by 11 feet deep, was carried away, along with about 150 feet of the main structure. The entire cap and the lower tiers of blocks, weighing 1,350 tons, were carried bodily away in one piece. The breakwater end was rebuilt and, in an effort to prevent a repetition of the destruction, a cap of 2,000 tons was made in order to tie the whole together. In spite of this, however, five years later the breakwater was again destroyed by another storm. Calculations showed that the titantic wave force necessary to achieve the second destruction was 6,340 pounds per square foot.

The breakwater at Cherbourg, France, was built of loose stone strengthened by 700-cubic-foot blocks and topped by a 20-foot-high wall. During a winter storm stones weighing nearly 4 tons each were thrown clear over the wall and some of the concrete blocks moved more than 50 feet. Even in such a comparatively small sea as the Mediterranean, with a limited fetch, enormous seas may crash against the coast. In February 1934, 30-foot waves engulfed the Mustapha jetty at Algiers. In this case the maximum fetch was 400 miles. In the comparatively restricted waters of the English Channel a 100-ton, fully laden sailing vessel was tossed upon Chesil Bank, on the coast of Suffolk, to a position 30 feet above high spring tides.

It must not be supposed that the damaging effect of waves breaking upon a coast is due only to local storms. An impressive example of the effect of distant storms producing a swell, the energy of which becomes concentrated in shallow waters, has been reported from Barbados in the West Indies. Between October 24 and 28, 1958, with no storms anywhere in the vicinity, heavy swells struck the northern coast. Although not especially noticeable at sea, these waves became shorter and higher as they entered shallow water, until 30-foot-high crests were crashing against the shore. Fishing boats were hurled upon the beaches and houses filled with water and sand as the seas invaded the land. There was no warning and no obvious reason for the onslaught. However, weather charts showed the presence of a large, intense extratropical cyclone in the North Atlantic in latitude 45 degrees north and longitude 44 degrees west, more than 2,000 miles away, between October 22 and 25. The time lapse between the peak of the storm and the arrival of the largest waves at Barbados was of the correct order of magnitude for the travel of swells of the wavelength measured. Systematic weather and wave forecasts now make it possible to predict the arrival of such waves from as much as 8,000 miles distance, with ample time to take precautions. The Barbados incident is by no means an isolated example. Fair weather waves of 40 feet are not uncommon on Indian Ocean shores, whose "rollers" are dreaded by the fishermen of Ceylon. The Moroccan coast is especially exposed to Atlantic swells from distant storms. In fact, most of the western shores of continents are commonly visited by huge fair weather waves, notably in South America.

The energy in a wave is proportional to the square of the height and to the wavelength.* Thus, a 5-foot-high wave of 10-second period has more than 100,000 foot-pounds of energy. If this amount of energy was expended over several seconds it would not be as destructive. But the effect is often concentrated into fractions of a second. The actual pressure excited at the very brief moment of impact has been estimated for a wave of 5-second period and 10-foot height at more than 1,200 pounds per square foot.

The design of breakwaters, jetties, and seawalls is a complicated matter that is beyond the scope of this book. Nevertheless, certain basic principles for designing harbor defenses should be evident from what has already been said, as the following examples may show. First of all, it is necessary to have a historical background of storm waves in order to determine the most frequent directions and probable maximum size of waves. Using charted depth contours, refraction diagrams may be prepared. From this it is possible to select the most suitable location, one that will provide the utmost protection and at the same time the minimum danger to the structure. For instance, it would normally be inadvisable to build a breakwater in such a way that the strongest waves approach from a head-on direction. Under such circumstances the whole force of the wave front would be expended upon

---

* Relation of wave energy (E) to wave length (L) and height (H):
$E = 8LH^2$

Relation of wave pressure (P) to wave speed (C), orbital velocity (V) and water density (ρ):

$$P = 1.31\,(C + V)^2\,\frac{\rho}{2}$$

the structure at once. If the location allowed the waves to approach somewhat obliquely, the forces would be expended more gradually. Since waves are reflected from a vertical wall, forming clapotis, this may be the best type of breakwater to build, the main energy being returned to the sea instead of the waves expending their force on the structure. However, this is only true if built on water sufficiently deep that the waves are not breaking.

In an exposed harbor, at an open dock structure, such as one built upon pilings, a ship is subjected to lateral surges to and from the dock, as swells pass through, with resulting strain on mooring lines. A solid structure will allow the development of clapotis, so that while the in and out movement of ships tied to the dock is lessened, the up and down movement is increased. A seawall may be designed with a concave outer wall, which deflects the breaking wave so that its energy is spent as a vertical gush of water that carries back into the sea. It may also be designed in such a way that the exposed surface is broken by irregular protuberances to dissipate energy. Another type of structure is built of receding steps, so that the impact of a wave is absorbed gradually, rather than in one great instantaneous blow. In all cases it is necessary to provide adequate drainage for water accumulating behind the exposed front, so that the structure does not become undermined by seepage.

In addition to breakwaters, there are a number of other devices designed to protect beaches from loss of material or the land from invasion by the sea, including various types of seawalls and groins. (The literature on the subject is considerable, so that readers must be referred to suitable books in the accompanying bibliography for detailed information.)

Nevertheless, the concrete tetrapod, a modern protective device, may be briefly mentioned both as an additional example of some principles involved and because it has become an object of curiosity on an increasing number of beaches. The tetrapod is a four-armed mass weighing from 1 to 50 tons, depending upon the local requirements. The stubby projections are spaced at equal angles. The general shape helps to dissipate energy, and at the same time, even though closely spaced together, the arms allow water to flow through freely, so that no great internal pressures are developed. The arms also interlock with adjacent tetrapods so as to provide mutual support. Experience has shown that these may be more effective than monolithic structures.

Diffraction of waves by an obstacle is greater for waves of larger wavelength than for waves of shorter length so that, just as in wave refraction, the obstacle, such as a harbor jetty, may provide better shelter from the shorter seas than from the larger swell.

The action of waves upon beaches lacks the spectacular fury of their impact upon large structures. Nevertheless, the results may be to remove large areas of beach or to build up shoal bottoms in harbors, and so necessitate the expenditure of huge sums to correct the situation or to replace buildings undermined by waves when the protecting beach is lost. When built, Montauk Point Lighthouse was 200 feet from the water. It is now separated from water by only 40 feet. There are innumerable other instances of the loss of land to the

One of many systems for beach protection is the use of concrete tetrapods. The four-armed masses may weigh from 1 to 50 tons. Their shape and size enables them to dissipate wave energy, but at the same time to allow free drainage of water and to provide mutual support by interlocking.

encroachment of the sea. On the other hand, there are also examples of the slow seaward extension of land that results from sediment deposition.

Simply stated, the effect of waves upon beaches is to move the sand, shingle, gravel, or other beach materials to and from the shore and also in a direction parallel to the shore. The to and fro movement is clearly demonstrated by the great difference in appearance of a beach between the stormy periods, usually in winter, and the periods of gentler waves, usually in the summer. In winter the beach is heavily cut back and steepened. In summer the beach materials are replenished and the gradient is lessened. The summer waves are smaller, with less energy. The sand is disturbed, lifted vertically and carried forward with the orbital movement of the wave and so deposited farther up the beach. The reverse movement is lessened because it happens in the trough and is impeded by friction. In this way there is a net shoreward movement. With large storm waves, however, the greater energy cuts deeply into the beach and keeps sand in suspension much longer. The seaward movement effects a removal of the sand, which cannot settle in the high energy zone of the breakers, although it may do so offshore, even to the point of building a bar. The beach contours also depend upon the particle size of the materials. For instance, a pebble or shingle beach is usually much steeper than a sandy beach. In constructing artificial beaches, it is necessary to provide an angle of slope that is stable under the conditions to be expected. The stable slope must extend into sufficiently deep water to withstand wave action.

Although alongshore water movements, already mentioned in connection with rip currents, may change seasonally, there is generally a net movement in one direction or another, depending upon the prevailing winds and the consequent angle at which the breaking crest meets the beach. The lateral water drift carries with it a greater or lesser amount of beach material, depending upon the strength of the wave action and its ability to keep particles in suspension. The significance of this is that beaches are not fixed, stable structures but are in a constant state of movement. Material is lost on the downstream side of the beach and is only replaced by material carried from upstream. If traced back along the coast, the original source will often be found in the sediment load supplied by a river. The continual supply of fresh material to a beach is usually referred to as nourishment. Convincing illustrations of alongshore drift are offered by such obstructions as groins, which are the familiar low walls that are built from the high tide line across the beach in a seaward direction and terminate some distance below the low tidemark. Any feature that prevents the natural drift of beach material will result in starvation and, eventually, loss of the beaches on the downdrift side. This can only be corrected, and even then only temporarily, by hauling sand for replenishment. It is true that sand accumulates on the upstream side of a groin since the area immediately adjacent to its upstream face is one in which there is relatively little energy and the materials are therefore deposited. The further drift of sand is prevented by the groin and any further progression must be beyond its seaward end. From here on it is liable to be deposited in deeper water and effectually lost if the groin is excessively long. Thus, the effect of a groin may be to protect and nourish the beach to windward but to deprive the beach immediately beyond it. If the groin is not excessively long, at least some of the material will be recaptured by the next groin below it.

A striking example of the dynamic nature of beaches and the results of interference is that of Lake Worth Inlet, Palm Beach, Florida. Here a jetty was built out to the north side of the inlet in order to prevent the prevailing southward movement of sand from entering the inlet and reducing its depth to the point where navigation would be restricted. The result was that a wide beach developed north of the jetty from the accumulated sand, whereas, south of the inlet the beaches, lacking nourishment, became impoverished. The remedy, an expensive one, was to install pumps to transport the sand through sunken pipes from above the jetty to the other side of the inlet.

A somewhat similar situation exists at Santa Barbara, California. The prevailing beach movement along the California coast is to the south and east. Below Point Conception the

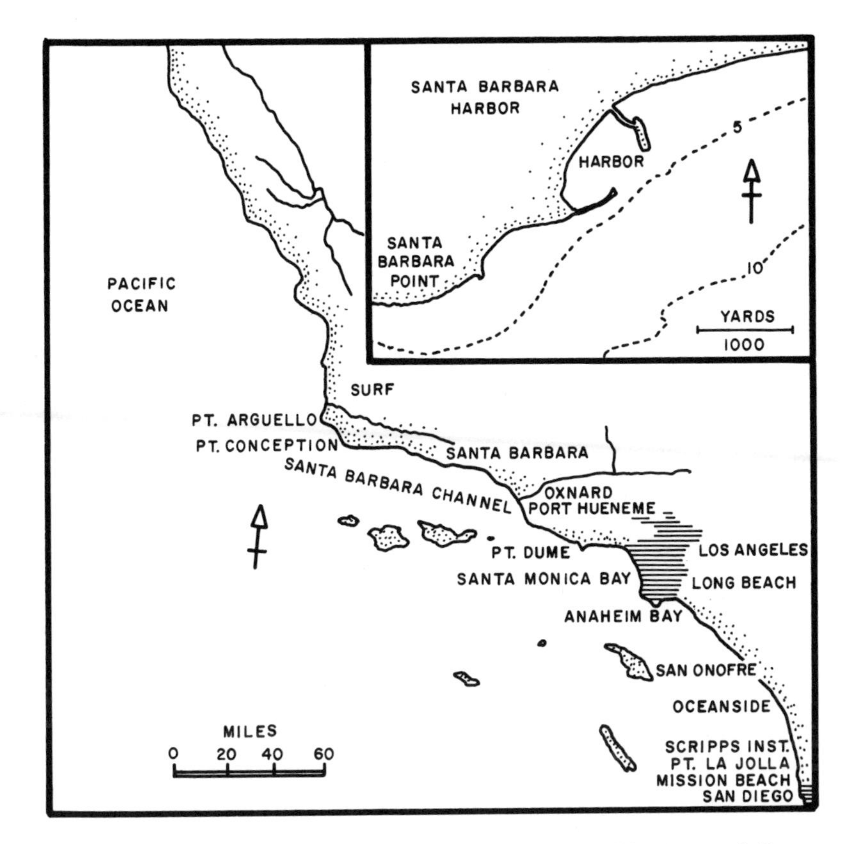

Coastline of California showing location of Santa Monica and Santa Barbara. At both places problems have arisen as a result of efforts to provide safe shelter for ships.

land curves easterly and the prevailing east-moving waves, though refracted, advance in an oblique direction to the coast. The amount of sand moving along the shore has been calculated at between 200,000 and 300,000 cubic yards a year, equivalent to a beach 1 mile long by 60 yards wide and 7 feet deep. At Santa Barbara a jetty was built out from an upstream point on the coast in order to reduce wave surges within the harbor. The result was that sand began to pile up to the upstream side and to accumulate inside the end of the jetty. The beaches for more than a mile on the downstream side, however, became impoverished and cut back more than 100 feet. The land itself was eroded and property was threatened. In this case the solution was to dredge sand from the shoaling area of the harbor and pump it to the starving beaches.

At Santa Monica, California, a somewhat different situation occurred. A breakwater was built offshore and parallel to the coast. The desired effect was accomplished, and boats moored behind the breakwater received a measure of protection. However, the reduction of wave energy also brought about an undesirable effect. Sand being transported alongshore began to settle in the shadow of the breakwater, where it was no longer held in suspension by wave action and thus could not continue its normal movement. Beyond the breakwater's shadow the beaches began to shrink from lack of nourishment.

At Long Beach, California, a costly lesson in the effects of wave refraction was demonstrated on two occasions. A breakwater had stood there for a number of years until, in April 1930, 18-foot waves over a period of four days displaced 20,000 tons of stone. The breakwater was repaired, but in 1939 it was again wrecked by wave action. A careful study showed that a long swell from a south-southeasterly direction became focused upon the harbor by the refractive effects of an 8-mile-wide shoal about 7 miles offshore. The waves were fo-

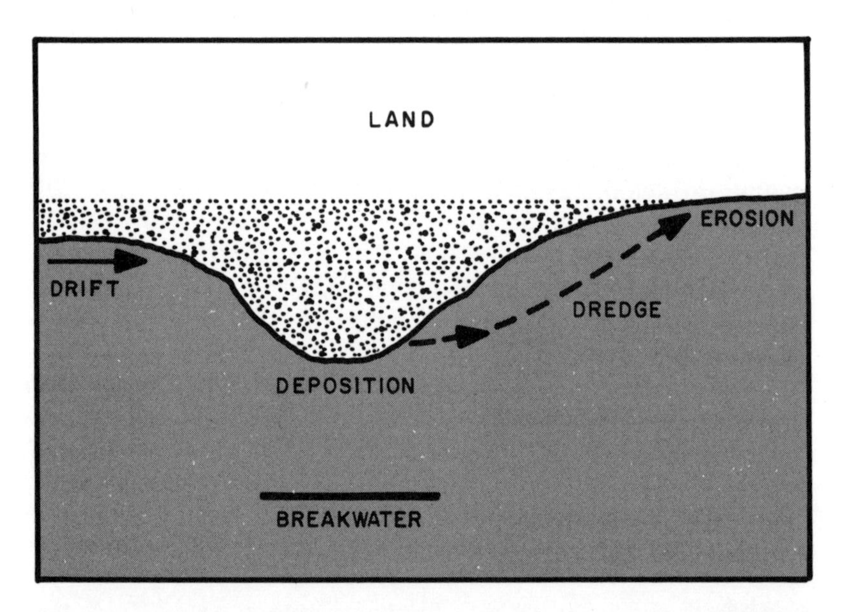

The harbor at Santa Monica was protected by a breakwater parallel to the shore. While this reduced wave action it also allowed sand drifting from left to right to become deposited behind the breakwater, and erosion to take place farther to the right.

cused onto the harbor structure, and a swell, calculated to be initially 2,000 feet in length with a 20-second period and a crest height of less than 4 feet, eventually became concentrated so as to develop 15-foot crests at the point of damage. A swell from a different direction would not have had the same effect.

To geologists, wave action is the architect of the coastline, taking advantage of the varying hardness of the rocks. A newly emerged coastline is heavily eroded where softer materials are exposed, forming embayments between outthrust forelands of harder rock. However, since wave refraction then begins to concentrate destructive energy on the heads and promontories, they begin to be cut back, materials are deposited in the bays, and the coastline begins to straighten out again.

Similarly, the combined effects of wave action and shore materials of varying nature are of interest to the biologist, because of the wide range of adaptations developed by the animals and plants that inhabit the zone where wave energy is expended. To mention extreme examples, a sandy beach under extremely calm conditions provides a fairly stable substrate for various burrowing creatures. Under conditions of heavy wave action, however, it is quite the reverse. In contrast, a rocky shore provides a firm surface for sedentary organisms. Those exposed to wave action need strong attachment mechanisms and a streamlined shape to resist wave drag, or else they are obliged to seek shelter in crevices. The action of waves also complicates intertidal effects, such as exposure to desiccation. The spray zone, for instance, is greatly extended by heavy wave action. Strong wave action also aids in transporting oxygen and nutrients and in carrying away waste products. Further consideration will be given to this in the chapters on tides.

Even the shell collector may be affected by waves. Storm waves, which disturb the bottom to a greater depth and to a greater distance offshore than shorter waves, deposit shells upon a beach that are otherwise not found frequently.

7. *More Waves*

In previous chapters attention has been paid to wind waves covering a spectrum of wavelengths, which begins with the fractional-inch capillary waves and includes ripples, chops, stormy seas, and finally the long swells that end as surf on coastlines. In addition to these, there are tides, with wavelengths extending halfway around the earth. There are also other waves, however, of various intermediate dimensions that arise from quite different sources of energy. They include internal waves, waves generated by moving ships, seismic waves caused by earth movements or volcanoes, microseisms, and waves caused by meteorological phenomena. All of these are governed by the general principles previously discussed.

Huge, unseen waves, slow and ponderous, are in motion below the surface of the sea. These internal waves differ from the familiar wind waves in being generated at a boundary between two liquids instead of the boundary between the sea and the atmosphere. The two liquids are horizontal layers of seawater of different densities, resulting from unequal temperatures, salinities, or both. Since the internal wave form is at some distance below the sea surface, it is not directly visible, although there may be indirect signs of its presence. Internal waves, because of the small density differences between the adjacent liquids, travel much more slowly than surface waves * but may be much greater in height, as much as several hundred feet in extreme cases.

---

* The speed of an internal wave. Where C is the wave speed, P_1 the density of the upper layer, P_2 the density of the lower layer, h_1 the thickness of the upper layer and h_2 the thickness of the lower layer, and h is the water depth. If both layers of water are deep, that is, more than half of the wavelength in depth, then the speed of an internal wave is given by:

$$C = \sqrt{1.56 \frac{P_2 - P_1}{P_2 + P_1}}$$

The reason for the increased amplitude is that where density differences at an interface are small less energy is required to displace the boundary. In fact, the amount of energy required to generate internal waves is several hundred times less than that needed to produce waves of the same height at the air-sea boundary. If a narrow glass dish several inches long containing water is rocked, the air-water waves generated will be of small length and amplitude, so that it is difficult to demonstrate breaking crests. On the other hand, if two non-mixable liquids of slightly different densities and colors are used, internal waves and the boundary between them are readily demonstrated and have sufficient amplitude to demonstrate the formation of surf. The desk-sized models demonstrating this action that are sold as novelties serve well to illustrate internal waves.

The boundary most commonly subjected to internal waves is the thermocline, where a rapid temperature change takes place. From the surface of the ocean down to this level the temperature is more or less the same. At the thermocline the temperature drops markedly. Below it, the temperature again drops, but more gradually with depth. The interface between the upper, warmer, less dense layer and the lower, colder, denser layer may be set in motion by various disturbing forces. A wind driving against the shore may tend to pile up the water so that the water surface near the shore is raised, for instance one inch above the general level. This will depress the denser layer of water below until the lesser amount of dense water balances for the added height of the lighter surface water. If the density difference between the two layers is in the order of one-tenth of one percent, then the denser water will be depressed by the amount of one inch divided

If the upper layer is shallow, then the wave speed is given by:

$$C = \sqrt{gh \frac{P_2 + P_1}{P_2}}$$

If both layers are shallow, then the wave speed is given by:

$$C = \sqrt{g \frac{h_1 h_2}{h_1 + h_2} \left(\frac{P_2 - P_1}{P_2} \right)}$$

by this amount, namely 1,000 inches, or about 80 feet. When the wind ceases and the surfacc recovers its previous position, the denser layer will return to its former level. In so doing it will generate an internal wave of approximately 80-foot amplitude.

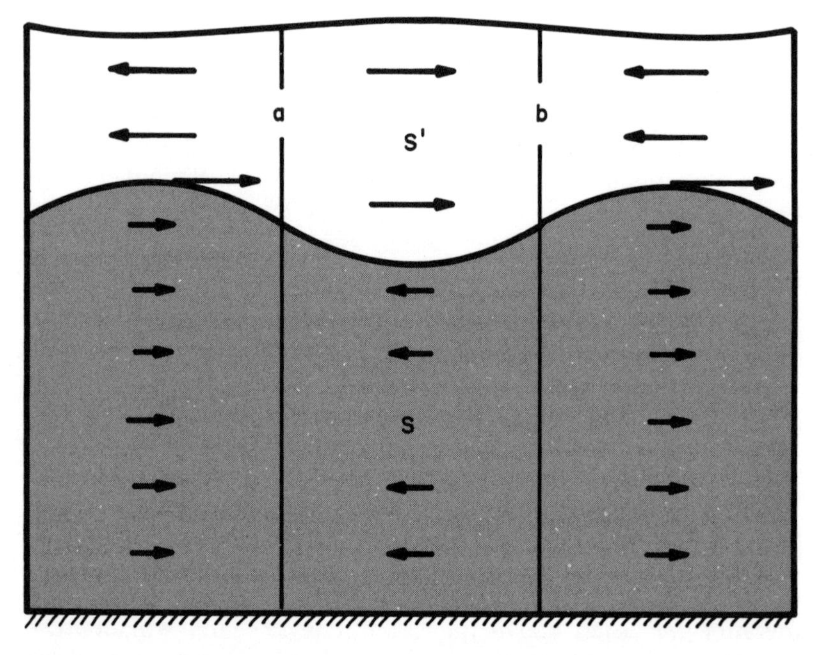

Diagram of water movements in an internal wave. The wave form is moving from left to right. At the crests, the water particles move to the right under the crest, to the left above it, with reverse movements at the trough. As a result, halfway between a trough and the next crest, water at the sea surface moves apart. Halfway between the crest and the next trough the surface water converges, often concentrating oil or debris into a "slick" line. S' indicates the less dense layer of surface water; S, the denser underlying layer. The surface convergence is above b and the divergence above a.

Internal waves are also generated when two adjacent currents of different densities travel in opposite directions or in the same direction at different speeds. The shear zone between the two bodies of water, frequently a thermocline, is thrown into a wave. An example of this occurs in the Straits of Gibraltar, where Atlantic water enters the Straits as a surface layer extending to a depth of about 300 feet and traveling at about one knot. For many years oceanographers were at a loss to explain what happened to this continual addition to the Mediterranean. Eventually it was discovered that an approximately equal but opposite current of denser, salty Mediterranean water flows out to the Atlantic below the thermocline. The instability at the boundary layer causes internal waves more than 60 feet in wave height. An especially interesting feature of these waves is that under certain conditions the crests break into a submarine surf, albeit in a slow and ponderous way. In the Strait of Messina internal waves and surf caused by the passage of lighter water from the Tyrrhenian Sea over that from the Ionian Sea cause disturbances at the surface, which in former times when the Strait was narrower and the eddies stronger may have given rise to Homer's description of the Charybdis whirlpool. Legend has it that the Strait was broadened in the 1783 earthquake, when the rocks of Scylla disappeared. An internal surf also forms where internal waves from the ocean break upon a continental shelf.

Although internal waves are not visible at the surface of the ocean, since the wave surface is below the sea surface, indirect evidence of internal waves may nevertheless sometimes be seen from a boat or aircraft. One indication of internal waves may be the appearance of elongated streaks that run in parallel lines at the surface of the sea. Internal waves generate movements on the water above them opposite to the orbital movements at the crest and trough of the internal wave itself. Thus, within the crest of the wave the water particles move in the direction of wave progression, but above the wave the water moves backward. At the trough the particles below the wave surface move backward but the water above moves forward. At the sea surface the result is an alternating series of bands parallel to the wave crests in which the water converges from both directions toward a band and diverges in both directions away from the adjacent band. Any detritus on the surface collects along the band of converging flow and changes the appearance of the water. If there is an abundance of plankton, producing oily materials, then the

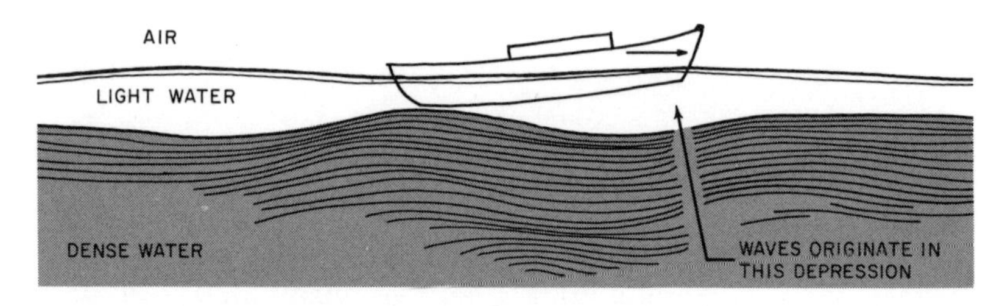

A ship may be slowed down in "dead water," which is caused by an internal wave developed by the ship's own movement.

band of converging water will appear as "slick." However, this is not the only cause of slicks.

Since internal waves are not directly visible, they were for many years undetected. How, then, are they now located and how are they measured, except by such signs as slicks or by their effects upon submarines? The answer lies in the change of density between the layers of water above and below the wave form. The change in density may be due to changes in salinity or temperature or both, but it most often occurs with a change of temperature. If the temperature is measured at a fixed depth at any one place, then the temperature recorded will change as the wave passes by, recording the lower temperature of the deeper layer of water as the crest goes by and the higher temperature of the upper layer as the trough goes by. The rate at which these changes occur is a measure of the wave velocity. Alternatively, an underwater buoy may be ballasted to remain in water of a fixed density at the location at which the greatest change is noted. The buoy will rise and fall upon the wave, and a recording of the changing depth will indicate the height of the wave from crest to trough.

Related to boundary waves is the phenomenon of "dead water." In Arctic seas it was noted that sailing ships moving slowly in light airs would sometimes move very sluggishly, as if restrained in some manner. If, however, they were able to pick up a speed of more than 3 or 4 knots, they might suddenly break free. It was eventually discovered that the phenomenon existed because of a surface layer of freshwater, meltwater from ice, overlaying the salt water to a depth of a little more than the draft of the ship. As the ship's bow pushes the surface water forward, the resulting pressure depresses the lower layer and starts an internal wave. Movement of the ship, therefore, is deprived not only of energy expended in its surface wake, but also in waves generated in the boundary layer beneath the keel. Because of the large amplitude of these waves, which have periods of 15 to 20 minutes and lengths of up to 1,000 feet, the ship is greatly impeded. However, since the waves travel at speeds below 3 knots, the ship needs only to sail at above this rate in order to escape. Similar situations exist in the Scandinavian fjords where freshwater runs out above the salt water.

An interesting relationship between internal waves and feast and famine in the herring industry was postulated by the Swedish scientist Otto Pettersson. From the thirteenth to fifteenth century a plentiful supply of fish regularly entered the fiords of Sweden and also the Baltic Sea through the sounds and belts from the North Sea. Since the fifteenth century, however, this bounty has dwindled. The cause for this may lie in the internal waves that enter these passages at approximately tidal intervals. Calculation shows that during the times of herring plenty the lunar and solar tidal forces were at a maximum; now, as part of a long cycle, those tidal forces have dwindled. The theory advanced is that the huge internal, tidally induced waves of yesterday swept the herring in front of them into the fjord, while the lesser waves of today are insufficient in amplitude to accomplish this.

Although not always of concern to surface vessels, internal waves may be hazardous to submarines. An underwater vessel, hove to and with tanks trimmed for neutral buoyancy, will rise and fall with boundary waves. If underway, however, the vessel could well move out of a water layer for which it is neutrally buoyant and pass through a large wave face into lighter water, thus causing a sudden dive. The result could be an involuntary descent into increased pressure depth. It has been conjectured that internal waves may have been involved in the mysterious loss of the U.S. submarine *Thresher*.

Internal waves are not always progressive. In the same manner as surface waves in an enclosed basin, they may become standing waves, analogous to clapotis, set into movement by tidal or wind influences so long as the waters are stratified in layers of different densities. There may also occur stationary internal waves, which are, in effect, progressive waves halted by opposing currents of velocity equal to that of the wave progression. They may be set in motion when a current in stratified water meets a bottom obstruction or is otherwise disturbed. Internal waves may behave as short waves if the water depth is large compared to the wavelength.

The movement of a ship through the water also causes disturbance, of course, and resulting waves. The energy used to propel a ship is consumed in various ways. Part of the energy is expended in overcoming the viscous resistance of the water, part of it is dissipated as eddies in the turbulence behind the stern, and part is used to generate waves that travel with the ship and cover a considerable area behind it. These waves are the most important to the ship designer in his quest for the economic use of power. Waves generated by a ship may be easily observed by towing a model hull in a swimming pool or by watching a swan or other water bird paddling across a smooth pond. Water flowing at a steady speed past an anchored floating object or fishing line will produce somewhat the same wave appearances as if the object were moving at the same pace through still water. These waves are of the same class as those produced when a stone is thrown into a pond and waves travel outward from the disturbance.

Ship waves in deep water, at their simplest, consist of bow and stern waves. Bow waves spread out behind the ship from the bow on both sides in a V. Behind the stern, parallel waves at right angles to the ship's direction move forward in step with the ship.

As the bow moves into the water ahead and disturbs it, the surface is lifted as a result of the pressure. This disturbance is propagated outward at an angle from both sides of the ship. An interesting feature of the bow wave is that as long as the ship continues to move the wave moves at a speed sufficient to keep pace with the ship. A simple diagram will explain this. When the ship reaches point *A* (page 94), it generates a wave on each side of the bow. As it moves on to *B* it continues to generate a wave at the point of the bow, but the part of the wave that was generated at *A* has now reached point *C*. At intermediate points the wave generated by the bow has reached various parts along the line *BC*. Thus the V-shaped wave lines stay with the ship. As the ship moves from *A* to *B*

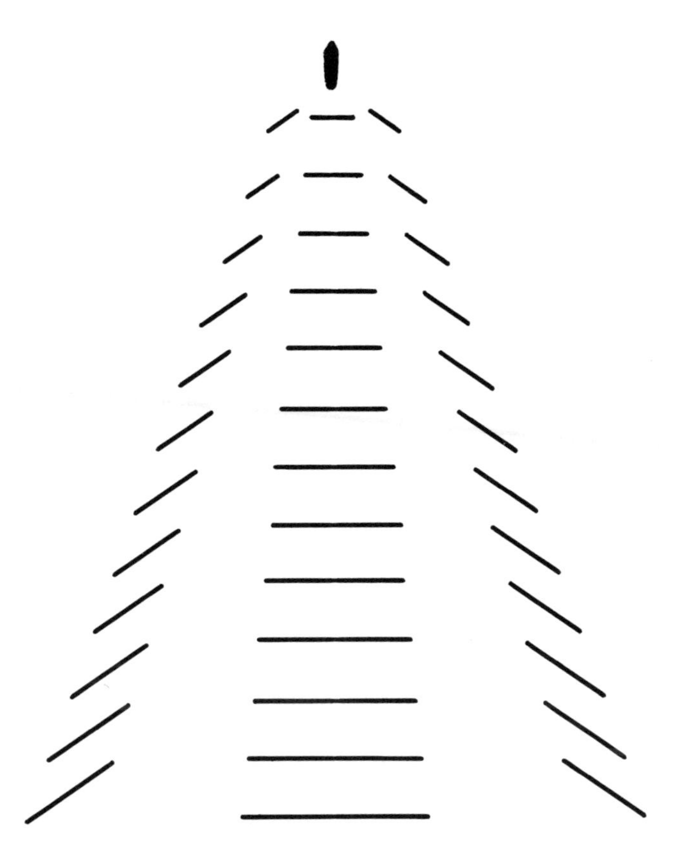

Bow and stern waves developed by the movement of a ship. The waves move at whatever speed the ship moves, so that the crests always maintain the same position and angle to the ship's course. If the ship speeds up, the waves also speed up, but the angle of the bow waves remains constant.

the wave moves from *A* to *BC*. If the ship were to stop suddenly at *B*, the wave would continue its motion, thus leaving the ship behind it.

Between the bow and stern the sea surface is depressed below the levels of the waterline when at rest, but at the stern a wave is formed at right angles to the ship's course. This is caused by the disturbance of the stern moving away from the water. The stern waves also keep pace with the ship. The waves produced by bow and stern are similar to the effect produced by the stone in the pond, in which waves of various lengths and speeds travel from the disturbance, but only those that are able to keep pace with the ship and reinforce each other remain in the system. The other waves travel out too fast or lag behind if moving too slowly and die down. As a result of this, the bow wave traveling with the ship always retains the same shape, namely a V, with the apex at the bow

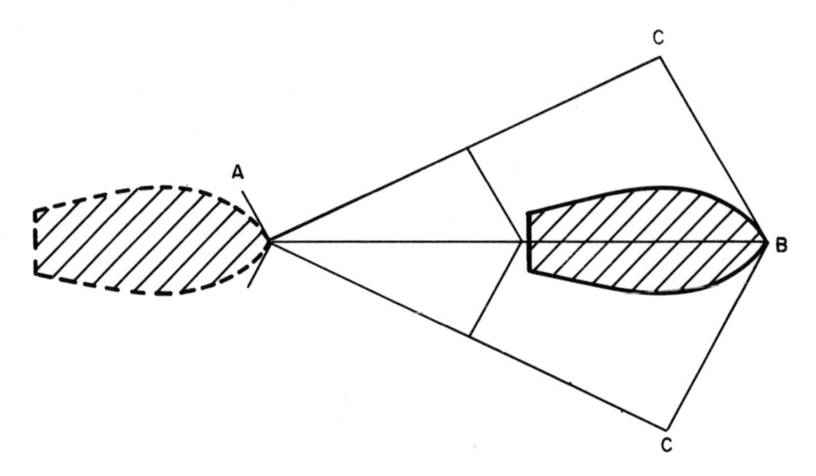

As a ship moves through the water, a bow wave is formed. This remains with the ship, but grows. If a ship starts at A, with the beginning of a bow wave, by the time it gets to B, the wave remains with it, but now has a crest extending from B to C.

and an angle between the two arms of the V of about 39 degrees, or about 19 degrees between each arm and the ship's course. No matter how slowly the ship travels this angle does not change. Another interesting feature of the V wave, derived from its origin in multiple waves that reinforce each other, is that each arm is not a single crest but is made up of a series of crests moving outward in a direction inclined at a constant angle of about 55 degrees to the direction of the ship.

If the ship slows down, although the wave pattern remains, the waves also slow down, the stern wave moving closer to the stern and the crests moving closer together as the wavelength shortens. If the ship stops suddenly, the waves simply overtake the ship and continue on their way, as mentioned previously. Conversely, if the ship speeds up, the waves also speed up, the crests move farther apart and the stern wave drops back to its new distance. This is because the length of the wave is proportional to the square root of its speed. Various differences in position of the bulge of the bow and stern waves are related to hull conformation and hull speed. Yet the wavelength is always dependent upon the ship's speed only and not upon its hull shape.

The fact that the wake waves keep pace with the ship makes it possible to calculate the speed of the ship. In deep water, where the speed of a wave is proportional to the square root of its wavelength, the ship's speed, which equals the wave speed, is calculated by estimating the length of the stern wake wave, multiplying by 1.8, and finding the square

5. SHIP'S SPEED AND LENGTH OF STERN WAVE

Ship's speed (knots)	4	6	8	10	12	15	18	20
Wave length (feet)	9	20	36	56	80	125	180	222

root. Thus, if the crests of the wake are 20 feet apart, the speed of the wave and of the ship is 6 knots ($\sqrt{1.8 \times 20}$). The relationship between wavelength and speed also makes it possible to calculate the most economical ship's speed because of the interference between bow and stern waves. If a crest of the bow wave coincides with the trough of the stern wave, they will tend to cancel out, and less energy will be expended, therefore, in generating the wake. On the other hand, if crests coincide with crests and troughs with troughs, the stern wave will be accentuated and more power will be expended. The maximum length of bow wave that will cancel out the stern wave is twice the length of the ship. The speed of wave for this length is $\sqrt{1.8 \times 2 \times L \textit{ ship}}$, giving a maximum economic speed of 8½ knots for a 20-foot ship, 12 knots for a 40-foot ship, and 20 knots for a 100-foot ship.

In large vessels, while the general features already mentioned for deep water still hold, the simple combination of bow and stern wave is made more complicated by the addition of other waves that may be generated from other parts of the ship, such as the shoulders and the quarters.

In shallow water, where the speed of the ship's wave is no longer dependent upon the ship's speed but upon the depth of the water alone, the behavior of the bow and stern waves change. This happens when the length of the wake wave becomes less than ½ of the depth. At this juncture, the speed of the wave is no longer dependent upon its length but upon the depth of water alone. If the ship is traveling at a speed less than or close to the critical speed, it becomes necessary for more energy to be exerted by the boat to break free of its wave, much as an airplane breaks through the airwave at the speed of sound. If the ship's speed is greater than this critical wave speed for the depth of the water, then the stern wave will disappear and the bow wave will become more and more parallel to the ship. Thus, if a ship has sufficient power to get up speed beyond the wave speed it will no longer be wasting as much power in making wave.

An interesting example regarding the importance of the critical wave speed in shallow water appears in an 1840 publication of the Royal Society of Edinburgh, in which the horse-drawn barges that plied the Glasgow and Androssan Canal were noted. Apparently a horse pulling the boat of William Houston became frightened and ran away, dragging the boat at an unusually high speed. The stern wave, which used to squander energy and destroy the canal banks, suddenly subsided, and the vessel was then drawn along smoothly but at a greater rate and with less effort. From then on it became the practice to whip the horse to a sudden jerk, which brought the boat to beyond the wave speed; the boats then moved at speeds of 7 or 8 miles an hour with less effort.

8. *Most Deadly Wave*

Intermediate-sized oscillations of the sea that can be extremely destructive are often called tidal waves, although they have no relation to the astronomically generated tides. They are caused, rather, by movements of the earth itself. Although scientists prefer to call them tsunamis, in order to distinguish them from true tidal waves, the Japanese term "tsunami" actually means "harbor wave." Probably the most appropriate name to apply is seismic wave, since this implies a wave resulting from some movement of the earth—a volcanic eruption or a massive landslide, either on the coast or underwater, or an earthquake. Any of these may cause a major disturbance within the sea, giving rise to a short series of very long waves.

One of the worst disasters caused by a tsunami took place at Hilo, Hawaii, on April 1, 1946. Because of the presence of a number of scientists who were visiting Hawaii at the time this was probably the best observed tsunami up to that date. Without any advance warning, the first huge crest broke over the north coast of Kauai and in one hour had swept around all of the Hawaiian Islands. In Hilo most of the houses on one side of the main street were picked up and smashed against those on the other side. Steel railroad bridges were torn off their foundations and tossed hundreds of yards. Fifty-nine people were killed and $25,000,000 of damage was done. Yet at sea, there was no sign of the approaching wave.

Investigation showed that at 2 A.M. a submarine earthquake had shaken the sea floor at a place near the Aleutians in Alaska, 2,300 miles from Hawaii. The origin was located at a point far below the surface of the ocean off the island of Unimak by seismographs. The quake was probably caused by a deepening of a fault trough in the sea floor into which the water had rushed. As the water overcompensated, a giant os-

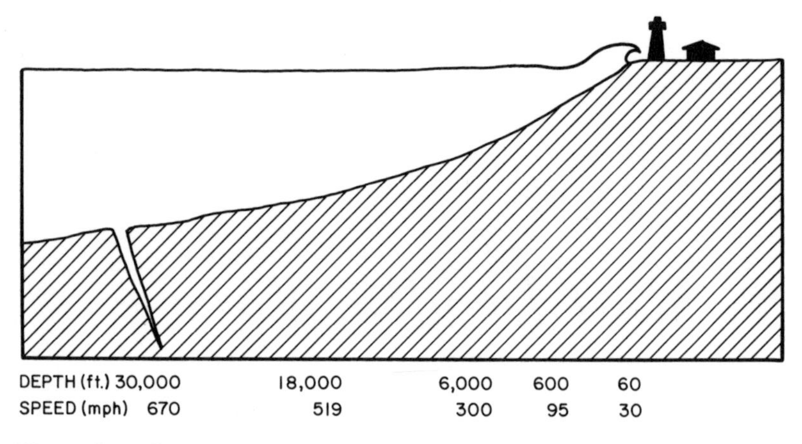

The path and speed of a tsunami. At the left of the diagram a disturbance of the sea floor, such as an earthquake or an eruption, causes a wave to develop. The wave is a long wave, with very low crests more than 100 miles apart. Waves of this length act as shallow-water waves and their speed is dependent upon the depth of water. In 30,000 feet of water they have a speed of 670 miles per hour. By the time they reach shore, in 30 feet of water, they have slowed down to 30 miles per hour. At this speed the crests are crowded together and become very much higher. Finally they may break over land, with a crest 100 feet high, and cause immense amounts of damage and loss of life.

cillation was set up and waves branched out at an average speed of 470 miles per hour. The fault trough in the sea floor, an area of weakness where part of the earth's crust has slid away from the neighboring rock, runs in an east–west direction so that the waves moved in greatest strength to north and south.

A little more than four hours later the wave had traveled as far as Hawaii. In the meantime the wave height, about two feet in the open ocean, was insignificant because the crests were more than 120 miles apart. Because of its great length compared to the depth of the water, the wave was of the shallow water type, the speed of which is proportional to the square root of the depth. Following the behavior of such waves, the speed and wavelength decreased as it entered shallower water near the coast, with a consequent increase in height. By the time it reached shore it broke in crests as much as 55 feet high at Polulu Valley and half as high at Hilo. Yet a captain on a ship lying offshore from Hilo did not notice the waves passing his ship, although he did observe them breaking onshore.

The 1946 tsunami consisted of several crests arriving about 15 minutes apart, which grew to a maximum height at about the fifth or sixth crest. A dangerous feature, which should have been a warning, was that the first easily noticeable evidence of the arrival of a tsunami, in an area where

The enormous power of a 100-foot wave was demonstrated at Scotch Cap, Alaska, in 1946. The reinforced concrete lighthouse shown in the upper photograph was utterly demolished. A radio mast, 103 feet above the sea, was also demolished by the wave, which was a tsunami generated by an Aleutian island earthquake. (U.S. Coast Guard)

tides rarely rise and fall more than 2 feet, was a withdrawal of water from the coast, like a rapidly falling tide sufficient to silence the usual uproar of the surf and to expose a large area of the sea floor. In other tsunamis at other places and times amazed and curious persons have wandered out to walk on the unexpectedly exposed sea floor only to be drowned as the crest arrived in a great breaking wave several minutes later.

The greatest waves of the 1946 earthquake were experienced at Scotch Cap, in the Aleutians, where a lighthouse stood on a foundation 32 feet above sea level, marking Unimak Pass, a few hundred miles from the source of the disturbance. On the morning after the quake airplanes were sent to investigate the lack of radio contact; the pilots found the merest traces of the foundations but no lighthouse, and the concrete base supporting the radio antenna, 103 feet above the water, was now bare. A wave estimated to be more than 100 feet high had swept over the light during the night hours. The five men who were on duty in the lighthouse, understandably, did not survive to tell of their experience.

Tsunamis, or rather their results, have been known from antiquity. The Greeks experienced giant waves, which must have been seismic in origin. Probably the first record was that of a wave that wiped out Amnisos in Crete about 1400 B.C. Although others have been recorded since then, only in more recent times, as communications and records improved and as population grew and spread to formerly isolated coasts, has the list grown more rapidly. Tsunamis have occurred most often in the northern and western Pacific and northern Atlantic, with the area a short distance south of the Aleutian Islands being the most frequent source.

Another area that has been a prime source for tsunamis is Chile in South America, which has tsunami records extending over hundreds of years. In 1575 the inner port of Valdivia was invaded by a giant wave that destroyed two Spanish galleons. In 1751 the town of Concepción was damaged by an earthquake and the sea receded and returned in several huge waves. The most amazing experience was that of the U.S.S. *Wateree* in 1868. On August 13 of that year, while the vessel was anchored at Arica, a tidal wave occurred that lowered the water in Iquique Bay by about 25 feet when it receded. This was followed by a great inrushing mass of water that covered the city of Iquique. The *Wateree* was carried by the wave right over the town and its buildings and was deposited almost a mile inland, miraculously without damage to her structure. Being flat-bottomed, she remained upright. While awaiting the sale of this landlocked ship her crew remained aboard. A garden was started

The U.S.S. Wateree was lifted on August 13, 1868, by a tsunami wave, which carried it clear above the city of Iquique and its buildings and deposited it a mile inland. (U.S. Navy)

on the adjacent land below her and, for obvious reasons, lavatories were erected "ashore," but naval routine was otherwise little disturbed. If the captain wished to go ashore the bosun's mate would pipe and issue the order "away brig." At this, the coxswain would run out on the boom, slide down a rope, untie a burro fastened at the bottom, and bring it alongside the ladder, which had been lengthened to extend to the ground. The captain would then mount and ride away over the dunes on his errand.

More recently, in 1960, Chile experienced a disastrous earthquake, accompanied by volcanic eruptions, extensive landslides, and other large earth movements. The immediate results over a 400-mile-long area were 4,000 people dead and $4,000,000 worth of property damage. But the most widespread effect came from a sudden slip of a great slab of the earth's crust on the large undersea fault running parallel to the coast. The result of this was a tsunami that radiated outward to long stretches of Pacific coastlines. Many coastal towns in Chile were devastated. As far afield as New Zealand and the Philippines, seaboard towns were flooded. Japan, nearly 9,000 miles away, was subjected to 15-foot waves, 180 people died, and damages amounted to about $50,000,000. In the United States, Los Angeles and San Diego also suffered damage to yachts and piers. The severest damage occurred, however, in Hawaii where even the havoc of the 1946 tsunami was surpassed. Fortunately, the population had been warned, and the loss of life was small.

A particularly severe wave hit the northeast coast of Japan on June 15, 1896. At the head of a bay the waves rose to the enormous height of 100 feet. In other parts of the bay waves ranged from 10 to 80 feet. Ten thousand houses were swept away, and, not surprisingly in the absence of warning, 27,000 people were killed along a coastal length of 150 miles.

Also in the Pacific Ocean and in the general area that has spawned many of the Pacific tsunamis was the great wave of Lituya Bay on the Alaska coast in July 1958. In this case the cause was not a submarine dislocation but a landslide caused by an earthquake. When the earthquake occurred, the gigantic shaking of the earth brought about landslides from 1,800 feet above sea level and triggered the fall of huge masses of ice from the fronts of two glaciers. The glaciers enter into the inner end of a steep-sided bay, in the center of which is Cenotaph Island. Across the entrance of the bay is a sandspit, which normally protects the haven of Anchorage Cove from the waves of the open sea. When the landslide took place and the huge masses of ice were thrown into the water, an enormous wave was started at the inner end of the bay. When the shock occurred, two fishing boats were anchored inside, one 40 feet in length and the other 50. The pilot of the smaller boat lived to tell of his experiences. He saw a great crest, more than 50 feet high, advance past Cenotaph Island, through Anchorage Cove, carrying his boat and himself high over the sandspit into the open sea. The boat, not surprisingly, foundered, but the fisherman and his wife escaped in a dinghy and were later picked up by another fishing boat. The 50-foot boat and its crew were lost.

In the Atlantic Ocean one of the first well-documented tsunamis happened in 1755, as a result of the great Lisbon earthquake that occurred on November 1. The waves that resulted rose to as much as 40 feet high along the coasts of Spain and Portugal. Smaller surges reached the shores of the West Indies, England, and North America. In 1929 an earthquake that originated on the rocks beneath the Grand

Banks caused a great wave that drove huge masses of water up several inlets of Newfoundland to heights that reached 50 feet above normal sea level, destroying several villages and causing heavy loss.

The long-distance record for tsunamis was set in 1883 when the volcano Krakatoa literally blew its top. The waves generated by this explosion left permanent records, as detected by scientific investigation, in the sea floor of oceans throughout the world. Since the end of World War II, oceanographers have intensified their studies of deep-sea sediments. Long pipes are dropped from research vessels and allowed to plunge into the soft ooze that covers the bottom of the open ocean in order to recover a sample of the accumulated deposit. The sample of bottom deposits is called a sediment core. Its vertical length is a history of what has happened to the ocean bottom, since the ooze accumulates at an average rate of an inch every 2,000 years. The volcanic dust that drops into the sea every time a major earthquake has occurred is interleaved into the layers of ooze, so that when the cores are examined, they are similar to the pages of a history book. Cores from almost all parts of the world show a white streak of volcanic dust bound into the accumulating layers of sediment at a depth below the present sea floor that corresponds to about the general time of 1880. This is the dust that dimmed the sunlight over much of the earth's surface when the volcano Krakatoa exploded on an August morning in 1883.

Krakatoa lies in the Sunda Strait between the islands of Java and Sumatra. It is a group of four islands covering an area of less than 20 square miles in a zone of heavy seismic activity, where three fissures of the earth's crust intersect. At one time Krakatoa was highly active, but it had gradually become, by the nineteenth century, an apparently moribund monster, though still, as it turned out later, with its fires only under temporary control. Before the great explosion occurred, an earthquake had disturbed the precarious equilibrium of the fissures and one of them was extended into the submerged crater, enabling seawater to flow into the intense heat source beneath the sea floor. Rumbling sounds became more frequent, and the Batavians who used to make

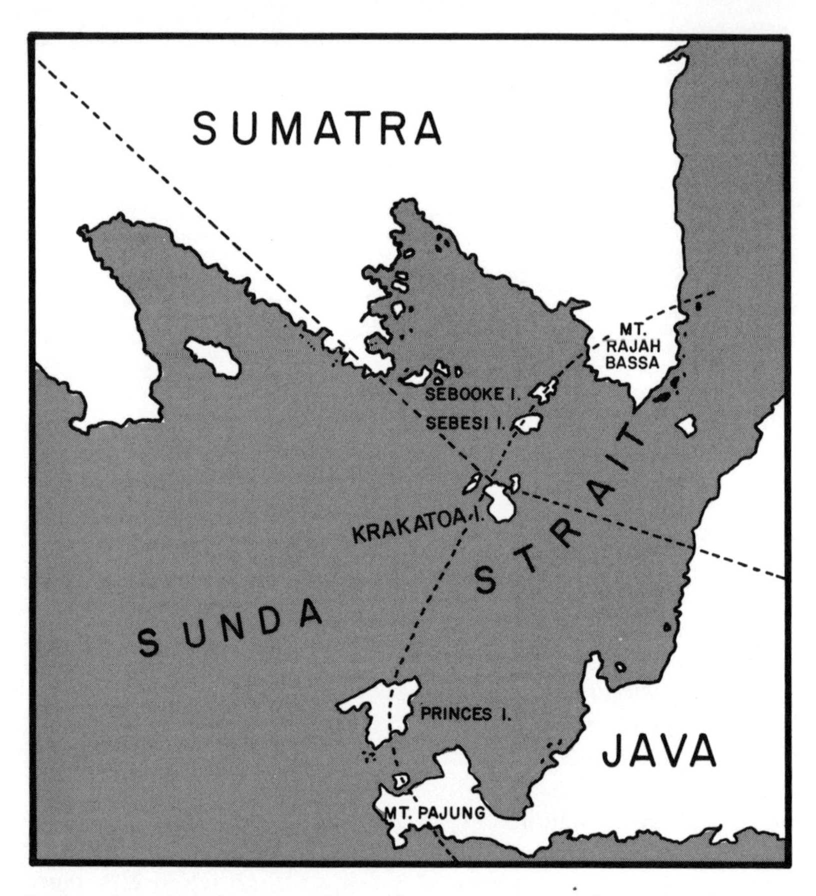

Broken lines in this map show the enormous earth cracks that converged at Krakatoa. At the point of convergence the sea poured in to meet the molten rocks beneath and caused one of the greatest explosions the world has known.

steamer excursions to the Krakatoa islands were finally obliged to give up these picnic pleasures by 1883. At that time volcanic activity had increased to an alarming extent, with rocks and pumice being thrown from the vent into the Strait, subterranean explosions growing in frequency and intensity, and a fine dust of volcanic material issuing from the volcano to form a somber cloud covering a wide area and visible at sea for hundreds of miles.

On Monday morning August 28 the rumblings gave place to a trio of gigantic explosions. The first explosion, early in the morning, was caused by the sinking of one of the islands, which had been 400 feet high, into a hole below the sea. Within about half an hour a second blast completely destroyed a second island, 1,500 feet high and much more extensive in area. Finally, with what has been described as the

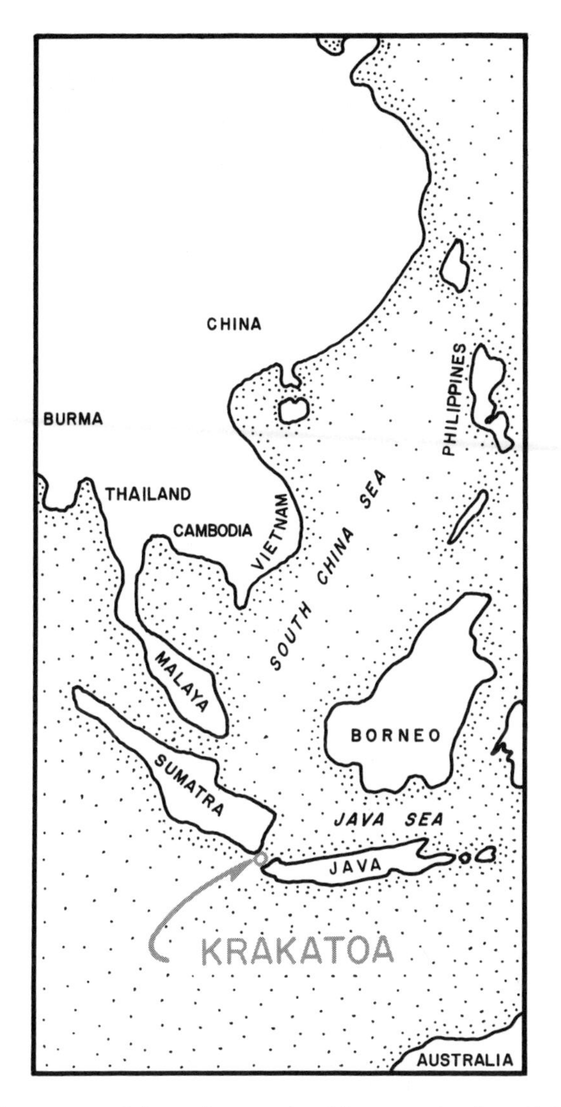

Between Sumatra and Java, in a zone of heavy seismic activity, is the group of islands called Krakatoa.

loudest sound on earth, an area of 5 square miles, including the remaining Krakatoa islands, was utterly demolished, leaving 1,000 foot depths where formerly three small mountains had stood, 399, 1,496, and 2,623 feet above the surface.

The tremendous force of the explosion, estimated as being equivalent to 20,000 megatons, vastly greater than any man-made nuclear explosion, was transferred to the ocean when it rushed into the enormous hole that replaced the islands. A giant wave spread out at a rate of 400 miles per hour, with its front a wall of water well over 100 feet high. The gigantic wave swept away a thousand or so villages on nearby islands,

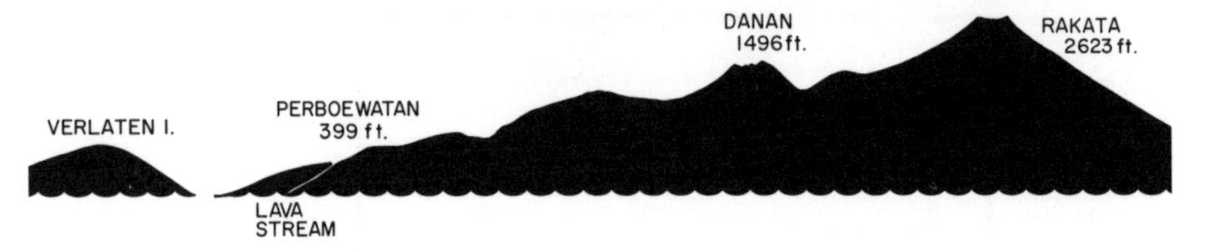

The three peaks of the island before blast are shown in silhouette. (Richard Marra)

where 36,000 lives were lost. Twenty minutes after the third explosion the harbor of Batavia was destroyed, along with every boat sheltering there. Nine hours later 300 riverboats were smashed at Calcutta. Altogether an estimated loss of shipping reached 6,000. Although ships on the open ocean were unaware of the wave passing beneath them, it ravaged harbor structures as far away as Perth in Australia. Halfway around the earth and 32 hours after the eruption, an admittedly lessened wave reached the English Channel and was recorded on the tide gauges. Only the astronomical tides can beat this speed and distance.

Associated with seismic waves is what was formerly a great mystery, that of the disappearing shoal. Ships would report a sudden shock, as if hitting an unchartered submerged reef, while crossing the deepest oceans. Subsequent passages by other ships, however, failed to show any trace of the underwater pinnacle. Such reports have been made by reliable navigators, who note that the ship's complement all agreed that there was either a sudden jolt or a sensation of bumping over a rough, hard surface. It is now believed that the solution to this mystery is to be found in shock waves generated by earthquakes. The shock wave is a compression wave in the rock of the sea floor that is transmitted vertically toward the sea surface at the speed of sound, about 3,500 miles per hour. It is of small amplitude, so that there is no visible disturbance of the surface to indicate its presence. Nevertheless, since the wave front hits the entire solid hull of the ship simultaneously, it has a shock effect. It may be compared to the effect of the sudden thump of a fist hitting a table. Although there may be no visible movement of the table, small objects lying upon it may jump sharply into the air.

One of the most recent reports of such an incident comes from Dr. Edward Hoffmeister, a Miami marine geologist, who was aboard the liner *Bergensfjord* en route from Lima, Peru, to Panama. On December 9, 1970, at 11:35 P.M. while in 600 fathoms of water, the ship was roughly shaken and a "loud, bumping, rasping sound" was heard, "as if it had hit

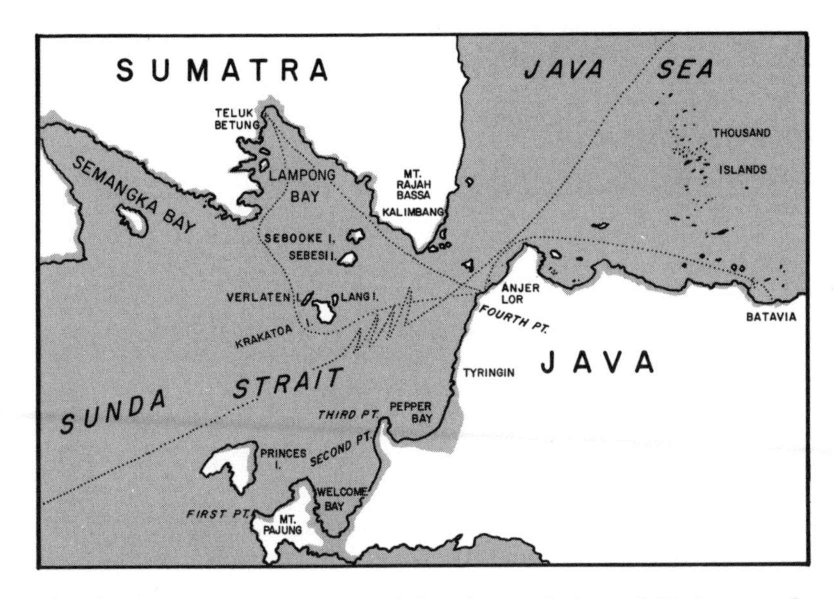

The devastating tsunami caused by the explosion of Krakatoa submerged and ravaged the areas shown in gray. More than 6,000 ships at anchor were destroyed. Yet, in the open sea, the waves passed under vessels, whose courses are shown by broken lines, without disturbing them.

something hard and was skidding over a rough surface." The ship listed slightly, but quickly righted herself. The engines were stopped and suitable precautionary actions were taken, as for a collision or grounding, but, since no damage was done the vessel soon proceeded on her way.

The next day a news release reported that a violent earthquake had racked northern Peru during the previous night, killing thirty-four persons. The ship's captain estimated that the ship's position at the time of its collision "with an unidentified object" was 70 miles west of the earthquake center.

The probability of sudden earth movements taking place along the rim of the Pacific Ocean, and thus releasing giant tsunamis to plague distant shores, is no less today than it was in 1946. The largest of oceans is ringed by a belt of great seismic activity, with major geologic faults, fractures, and deep, unstable trenches of island areas, which have earned the Pacific Ocean borders the name of "Ring of Fire." But, today, the giant wave is being controlled. The huge walls of water may still invade the land, but instead of thousands dying, there may be only tens. Since the 1946 disaster, scientists have worked to develop a warning service that enables the people of endangered coasts to take shelter from the expected waves and to save their lives, even though their property may still be damaged. The Seismic Sea Wave Warning

System was put into operation in 1948. In 1952 a submarine earthquake near the Kamchatka Peninsula threw out a wave that caused $800,000 damage in Hawaii—but it took no lives.

The earliest possible warning, of course, is one that is based upon immediate detection of the seismic cause rather than the wave in its subsequent travel. Because of the great speed at which a tsunami travels, every minute of warning is important; the first link in the warning system consists, therefore, of a network of seismograph stations. When an earthquake of sufficient intensity to generate a tsunami occurs, each seismograph station is alerted by an automatically triggered alarm.

The seismograph is an instrument that responds to the vibrations caused by violent earth movements, even at great distances. The vibrations traveling through the earth from the disturbing origin are picked up and transformed into electrical signals that deflect a recording arm so as to leave a record upon a moving band of paper. The record, called a seismogram, enables seismologists to determine within certain limitations the magnitude of the earthquake and its surface distance away from the instrument. Two kinds of waves are recorded. One is the P wave, which is longitudinal and vibrates in the direction of travel, the other is the S wave, which oscillates in a transverse direction. Since these travel at different speeds, the interval between their times of arrival at the station makes it possible to compute the distance of the source. The intersection of distance arcs drawn from several stations pinpoints with considerable accuracy the earthquake location. If this point, the epicenter, is on or near the ocean the possibility of the birth of a tsunami exists. An advisory is then sent out to all vulnerable points in the system, giving the approximate time at which the first wave might be expected, if it were generated. This time schedule is based upon a knowledge of the ocean depths along the line of propagation, since the wave speed is dependent solely upon depth of water.

A second link in the warning system consists of a network of tide stations, the records of which indicate the abnormal oscillations of a passing tsunami. Special sea level gauges that respond only to the long wave oscillations of a tsunami and ignore ordinary short wind waves are used. Reports from these stations to the central Seismic Sea Wave Warning System enable a second warning to be issued that a wave has, in fact, been generated. The first warning was of a possibility, the second of a strong probability. The few hours of advance notice have turned sudden disasters into avoidable death.

The present degree of understanding of the enormously complicated nature of the sea surface could not have been reached without means of measuring the movements not only of tsunamis, but also wind waves and tides. Some of the instruments and methods have been briefly mentioned in earlier chapters. There are a number of such devices, operating on different principles, some being suitable only for shorter wind waves, some for the longer waves such as tides and tsunamis, some for internal waves. Others are suitable for measuring the whole spectrum of waves with certain modifications.

Great advances in the design of measuring devices have been made in recent years. Space does not permit, however, a detailed description of them all; a brief explanation of the main principles will nevertheless give some idea of the ingenuity that has been applied to the complex study of the restless surface of the sea.

A surface float connected by a suitable mechanical link to a recorder can measure wind waves if the float is located immediately upon the open water or in a well connected to the open water by a wide pipe. The pipe must be wide enough in relation to the cross section of the well to allow the transmission of enough water during the passage of the wave to bring the water level in the well to that of the wave crest. If the system is used for tides alone, the pipe must be sufficiently small to prevent any perceptible change from taking place in the well level as a wave passes and yet large enough to allow a flow of water from the slow tidal change in sea level to take place. This is analogous to the filter system used to separate waves of different lengths in analyzing a wave record.

The simplest device for measuring wind waves consists of

a staff fixed to the bottom and extending perpendicularly above the surface. The graduated marks on the staff are read visually, using binoculars if necessary, to observe the wave height or are used with a stopwatch to determine the period. A more sophisticated version is provided with a high resistance electric wire running from top to bottom. One terminal of a recording resistance meter onshore is connected by cable to the top of the wire and the other terminal is allowed to float in the water. As the water rises and falls it short-circuits more or less of the resistance wire. The fluctuating resistance, properly calibrated, records wave height and period. Another type of wave staff uses an insulated cable dipping vertically into the sea. Changing wave height varies the electrical capacity between the insulated conductor and the water. A recording device onshore records wave height and period.

Still another system makes use of a narrow beam recording echo sounder, inverted in a fixed position beneath the surface. Just as a depth sounder on a ship measures the distance of the sea floor from the hull, so the echo sounder measures the distance of the sea surface above the sounder. In deep water a submarine may be used to support the echo sounder. Conversely, the echo sounder can be mounted on a floating buoy, so as to record the changing depth beneath it resulting from wave movement.

Because the surface of the sea is usually not a simple wave train but a series of short crests, hills, and hollows, two cameras fixed at a distance apart vertically above the surface may be used to record stereophotographs, or moving pictures of the surface. The sea surface profile may also be rapidly recorded by a low flying airplane using a radio altimeter.

The devices so far discussed are designed to measure the changes in sea level resulting from wave action. Other devices are designed to measure the pressure changes beneath the surface caused by passing waves. These are not altogether satisfactory, since they do not measure the actual weight of the water above them unless they are very close to the surface. At a depth of one-fifth of the wavelength, for instance, the variations are only one-half of the actual variations in static head. In measuring tides, however, if placed more than the wavelength of wind waves beneath the surface, there will be only negligible effects caused by the waves and only the slow tidal movements will be recorded. This type of instrument is essentially a rubber bag under the water connected by a pipe to a pressure recorder. If it is desired to measure simultaneously both wind waves and tides the same principle may be applied as in the well and float system. The pressure pipe is connected directly to the wave recorder, and through a narrow capillary to the tide recorder. The more rapid wind wave movements do not pass the capillary. These pressure changes may operate a bellows, similar to those that record pressures in aneroid barometers, to drive either needles on dials or pens recording measurements on moving paper. Another type of pressure gauge uses piezoelectric crystals, which develop an electric potential when subjected to pressure, to sense pressure changes.

An entirely different type of instrument, the accelerometer, measures the up and down accelerations of a buoy or ship by means of a weight suspended by a spring. An electric integration circuit converts the vertical accelerations into the actual distances moved. When combined with a pressure gauge in the hull of a ship, the accelerometer gives the changing height of the water. This type of instrument is extremely useful and has been refined to the point where it can be both accurate and reliable.

Pressure gauges of various types may be mounted on buoys anchored in a fixed position below the surface by a cable. They may also be suspended from a buoy at a depth below wave action. As the buoy rises and falls the gauge measures the changes of pressure as it moves vertically in the still water.

9. *The Wave of the Tides*

Far longer and faster than tsunamis are the tidal waves, with crests as far apart as the earth's circumference and with speeds that would keep pace with the rotation of the earth if no land intervened to impede them. On an earth completely covered with water, tidal waves would travel along the equator at about 1,000 miles an hour, or ⅓ mile per second. Like the tsunami, the tidal wave is unnoticed in the open ocean, revealing itself only by changes in sea level as it approaches land. While in some places the tidal rise and fall may be 50 feet or more, in others it is almost negligible. More important than the size of the rise and fall, however, is that the tides occur, regardless of size, in regular and predictable cycles.

Although tides have a regularity of their own, they do not occur at the same time each day, nor do they rise and fall by equal amounts at all times. Periods of small tides alternate with periods of large tides, and especially large tides may occur at certain times that may be years apart. Although tidal waves are independent of wind and weather, meteorological effects and the effects of other wave action may combine with tidal effects to bring about exceptional conditions. The coincidence of high tides with tsunamis or hurricane waves can be disastrous.

While tides in the open ocean are of little concern except to scientific investigators, the tides of coastal waters and estuaries are of concern to many people, not only because of their rise and fall but also because of the currents associated with them. In some places they are of sufficient magnitude to bring navigation to a halt or turn harmless waves into the steep breaking waves of a tidal race. Commercial fishermen, anglers, divers, shell collectors, and biologists as well as sailors have a need to understand and to predict tides.

Tidal movements, ocean currents, and waves are all manifestations of energy, derived either from solar radiation or as a result of the rotation of the ocean planet. Attempts to harness this ocean energy for the use of mankind have been largely restricted to the energy of tides. Even today, however, this energy source is insignificant when compared to other available energy sources.

Although the shorter wind waves, as well as tides, have been objects of concern to mankind from remote antiquity, they were less subject to philosophical inquiry than tides, since they came and went with seemingly no predictable cause. Indeed, some of the most destructive wind waves could arrive on a calm, peaceful day with no storms in the offing. But the regularity of tides was noted in the earliest writings and many intriguing theories were advanced to explain them.

The earliest theories as to the cause of tides are buried in mythology. Some ancient Chinese believed that water was the blood of the earth and that tides were the beating of its pulse; others believed that tides were the breathing movements of the earth. The ancient Mediterranean civilizations did not contribute much to explaining the cause of tides, mainly because the tidal ranges in that sea are rarely more than a foot; attention was more likely to be given to disastrous inroads of the sea brought about by storm surges or earthquakes, as indicated by the persistent legends of the deluge and the disappearance of Atlantis. The intellectually superior Greeks and Romans nevertheless did, of course, develop schools of philosophers who laid the early foundations of astronomy. Although the tidal information upon which they exercised their intellect must have been minimal at first, their information gradually increased by word of mouth as the Phoenician, Greek, and Hebrew mariners ventured beyond the Pillars of Hercules into the Atlantic Ocean, to the isles of Great Britain and to the North Sea coasts of the continent. To a Mediterranean seaman, the 30-foot tides of the French coasts must have been a formidable introduction to the gravitational forces of the sun and the moon.

In 350 B.C. Aristotle reported, "It is even said that many ebbings and risings of the sea always come round with the moon and upon certain fixed days." Pytheas, both philosopher and navigator, observed some years later, when voyaging from Marseilles to Britain, that tidal ranges differed at springs and neaps. Yet Julius Caesar was naïve enough to haul his galleys upon British shores during neap tides only to have them destroyed with the arrival of spring tides.

Pliny summarized much of the accumulated tidal knowledge in his *Historia Naturalis* in the first century A.D. He recorded the time intervals between the full moon and the occurrence of high water and noted that spring tides tend to be greater at the equinoxes, when the sun is at the equator, than at the solstices.

Much of the tidal lore became lost, except perhaps to fishermen, during the Middle Ages, so that the fate that befell King John of England in 1216 is not altogether surprising. The English history books relate that during a march from Lynn toward Wisbech, he attempted to ford a river but was surprised by the flood tide. He lost part of his army and all of his baggage and valuables in the waters of the Wash.

With increasing knowledge of astronomy, from the seventeenth century onward, more and more relationships were noted between the variations of the tides and the rhythmic movements of the moon and the sun. Even though empirical relationships were established between the heights and times of tides and the phases of the moon, however, the underlying causes still remained to be discovered. It was Sir Isaac Newton who finally extended his gravitational theory to reveal the forces that govern tides.

Newton's theory, the equilibrium theory, made clear that the complicated rhythmic cycles of the tides were governed by the gravitational attractions of the sun and the moon upon a rotating earth. Although the general nature of tidal rhythms was elegantly explained by the theory, questions of local pecularities were still left unanswered, since the equilibrium theory begins with a simplified situation in which it is assumed that the earth is completely covered by water with no land masses to interrupt the free passage of the tidal wave.

The gravitational attraction exerted by the moon upon the earth causes, as it were, a bulge of water to form directly under the moon. The earth rotates on its own axis, but since the bulge remains under the moon, any given place on earth will pass underneath the bulge once every complete rotation. It is not immediately apparent, however, why gravitational attraction also causes a bulge to form on the side of the earth directly opposite that under the moon, the moon's nadir. This may be simply demonstrated by attaching a rubber balloon filled with water to a cord held in the hand. If the balloon is now swung in a circle the balloon will be seen to stretch so that bulges form at each end of it and distort the original sphere into an egg shape. The distortion is a result of two opposing forces. The balloon is prevented from flying off at a tangent by the tension of the cord held in the hand. This centripetal force, so-called because it is directed toward the center of the circle, balances the centrifugal force, which is directed outward.

In the case of the moon, the centripetal force is the gravitational pull exerted mutually upon each other by the earth and the moon. If this pull were unopposed, the two bodies, the earth and the moon, would move toward each other. However, there is a centrifugal force that holds the system in balance. This is due to the fact that earth and moon revolve about a

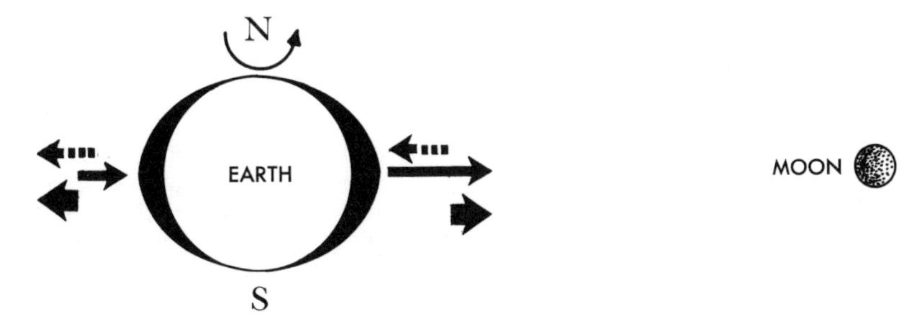

The forces responsible for tides are made up of the gravitational attraction of the moon for the earth (lighter unbroken arrows) and the centrifugal force resulting from the revolution of earth and moon around a common center (broken arrows). The centrifugal force is the same at all points on the earth's surface, but the gravitational force is greater on the side nearest the moon. The total effect is a balance toward the moon on the nearer side of the earth and away from it on the opposite side (heavy arrows). (Richard Marra)

common center, much as the balloon revolves around the hand.

The tidal forces of the earth have been calculated. They are the result of the centrifugal and centripetal forces and are inversely proportional to the cube of the distance between earth and moon.*

As mentioned above, the centrifugal force in the case of the earth exists because the earth and the moon revolve about a common center. This force is proportional to the square of the rate of revolution and inversely proportional to the distance between the center of rotation and the center of the earth. It is the same at all points on the earth. The gravitational force is inversely proportional to the square of the distance separating any place on earth from the center of the moon. It is therefore greater on the side of the earth nearest to the moon and less on the side farthest away. The difference between the two forces is the tidal force. At the center of the earth both forces balance. However, at the point on the earth nearest the moon the difference between the two forces is a pull toward the moon. On the opposite side the difference is a pull away from the moon. Hence, there are two bulges and potentially two tides a day.

The tidal forces acting directly beneath the moon are vertical. They are equal to only 1/10,000,000 of the earth's

---

* The forces act vertically upward from the surface of the earth at zenith and nadir, and at those places the force on unit mass is equal to $2agM/D^3$, where D is the distance between the centers of earth and moon, M is the mass of the moon, g is the gravitational constant, and a is the radius of the earth. At points midway between zenith and nadir the force is directed toward the center of the earth and is equal to agM/D^3.

The centrifugal force acting on unit mass at any point on earth is equal to $\frac{V^2}{r}$, where r is the distance between the center of the earth and the common center of revolution of earth and moon, and V is the speed of revolution. At a point Θ in longitude from the zenith, the centripetal force is gM/R^2 where R is the distance from the moon. But R is approximately equal to $D - a \text{ Cos } \Theta)^2$. The centrifugal force is GM/D^2. The horizontal components of these will be $gM \text{ Sin } \Theta/D^2$ and $gM \text{ Sin } (\Theta + \epsilon)/(D - a \text{ Cos } \Theta)^2$ where ϵ is the angular distance of the point from the line joining the centers of moon and earth.

The resultant horizontal tidal force is therefore $gM \text{ Sin } (\Theta + \epsilon)/(D - \text{Cos } \Theta)^2 - gM \text{ Sin } \Theta/D^2$.

This becomes $3/2 \cdot gM/D^3 \cdot \text{Sin} 2\Theta$.

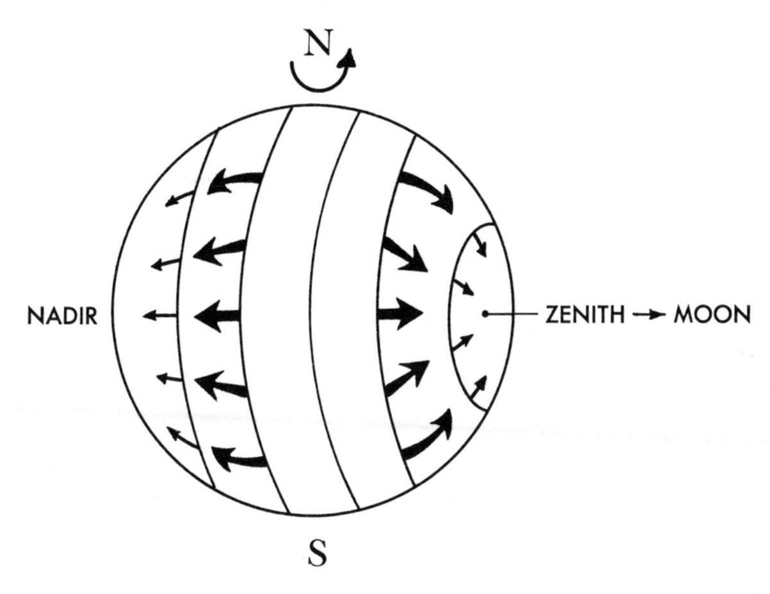

The vertical forces acting against the earth's gravity have little tidal effect. The horizontal components of these forces, moving the water across the surface of the earth, combine to pile it up toward the zenith and nadir. (Richard Marra)

gravity and would not decrease the weight of a grown man by more than the weight of a drop of water. Thus, by themselves, they would produce a tide too small to measure. At other points on the earth's surface, the tidal forces are different in direction and strength. At points halfway (90 degrees) between zenith and nadir, they are directed downward toward the center of the earth. At intermediate points, the tidal forces are directed more nearly parallel to the earth's surface.

The combined effect of the tidal forces operating over the whole earth is a tendency to pull it into an ellipsoid or egg-shape, with bulges beneath and opposite the moon. As a matter of fact, the solid rock of the earth actually does become slightly deformed by the tidal forces. But the most obvious result is that the waters of the earth, which are free to move, are drawn up into the crests of a wave, which is drawn across the earth as it rotates beneath the moon.

Any place upon the rotating earth passes under the moon once every day. Since the moon is also moving around the earth, the actual period between moon transits is not 24 hours, but 24 hours 50 minutes. Thus, at any one place on earth, high tide will occur every 12 hours 25 minutes. These tides are the semidiurnal tides, the morning and evening tides.

The sun also influences the tides, in a manner somewhat

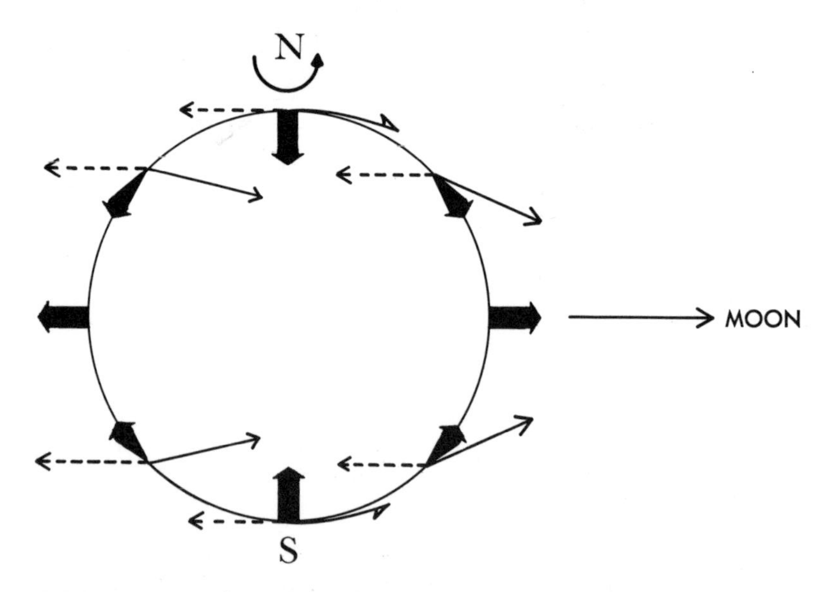

Tidal forces, resulting from the moon's attraction (light arrows) and the centrifugal force of revolution (broken arrows) are shown here (heavy arrows). At the points beneath and most remote from the moon (zenith and nadir), there is a vertically upward force. At points midway between these, the force is vertically downward. At intermediate points, the force is directed in a more nearly horizontal direction. (Richard Marra)

similar to that of the moon. Although the sun is considerably larger than the moon—its mass being about 27,000,000 times greater—it is also 389 times more distant from the earth—92,900,000 miles away. Since the tidal forces are inversely proportional to the cube of the distance between the earth and the moon and sun, the ratio of the sun's tidal force to the moon's is thus $27{,}000{,}000/389^3$, which is 0.46 or less than half that of the moon.

At times of new moon and full moon, the earth, moon, and sun are in line, so that the tidal effects of the moon and sun combine and become 1.46 that of the moon alone. This happens at intervals of about two weeks and the resulting tides, with a greater range, or amplitude, between high and low, are known as spring tides. At the times of the moon's first quarter and third quarter, the moon and the sun are at right angles in relation to the earth and their tides tend to cancel out. The result is that midway between the spring

When the moon is in line with the sun, the tidal effects of both are combined to give tides greater than usual, known as spring tides. When the moon is between sun and earth, at new moon, it is said to be in conjunction. When on the opposite side of the earth to the sun, at full moon, it is said to be in opposition. (Richard Marra)

tides, the tidal range is at a minimum. These are neap tides, at which time the force is 0.46 less than that of the moon alone, or 0.54 of the moon. So, the spring tide force is 1.46/0.54, or nearly three times that of the neap tides. The time between new moons is 29½ days, and this is therefore the interval in which two sets of spring tides appear: the new-moon spring tides and the full-moon spring tides.

In addition to the various rhythms, delayed actions in the tidal rhythms also occur. The tides may get out of step with the moon. Some tides occur before the transit, or passage, of

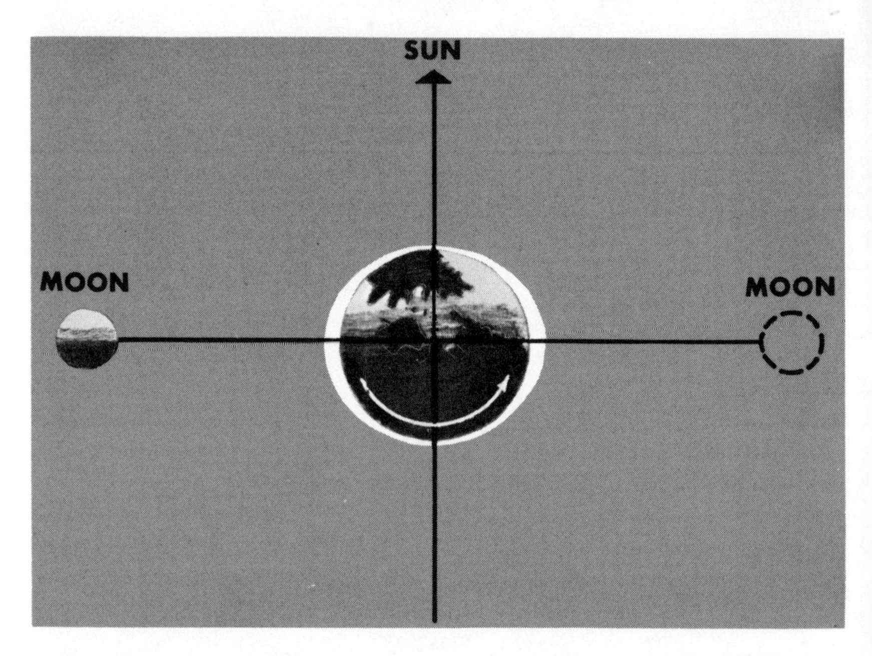

When the moon is in quadrature, at right angles to the line between sun and earth, its tides are partly canceled out by those of the sun, and smaller tides result. These are the neap tides. (Richard Marra)

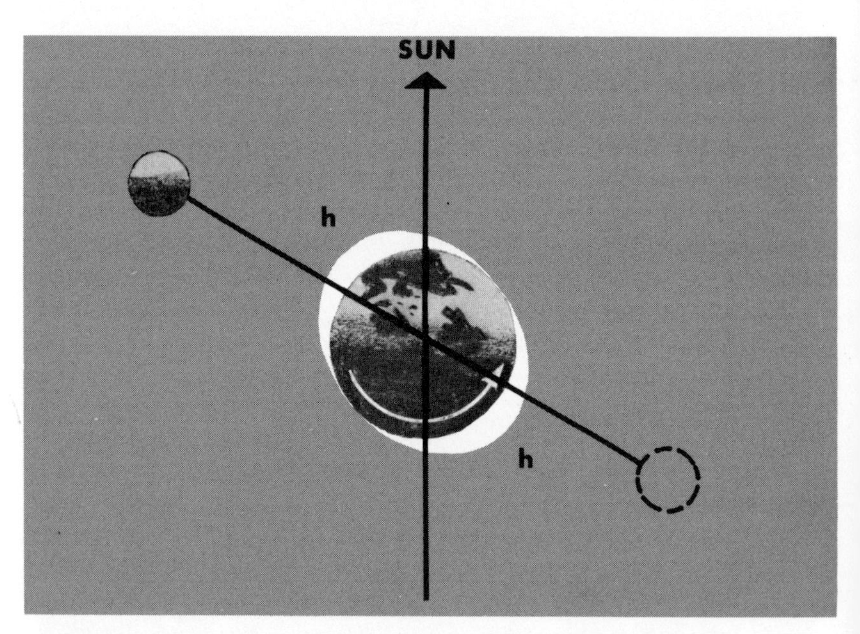

When the moon is in its first or third quarter, as shown here, the effect of the sun's gravity is to pull the tide slightly away from the moon. As the earth rotates, any particular spot will arrive at the tide before arriving beneath the moon. These early tides are said to be priming. In the second and last quarter, the tides are delayed and are said to lag. (Richard Marra)

the moon across the meridian. Others occur after the transit. These variations in timing are known as priming and lagging, and are partly the result of the relative positions of the heavenly bodies. When the moon is full or new, the sun and the moon are in line and so reinforce each other. High tide then occurs as the moon appears overhead. When the moon is in its quarter, however, the moon and sun are out of step and the resultant tide is displaced toward the sun. The effect is that when the moon approaches the first quarter high water comes before the moon crosses the meridian. In the last quarter, high water arrives after the moon reaches its zenith.

A feature that has not been explained to general satisfaction is that the highest of the spring tides arrives a day or two later than a full or new moon. Neap tides similarly arrive later than quadrature. This delay is known as the Age of the Tide.

Other variations in tides are caused by the fluctuating distance between the moon and the earth. When the moon is at perigee, or closest to the earth, it is 222,000 miles away and the tidal force is 22 percent more than average. When the distance is greatest, at apogee, about 253,000 miles away, the force is 16 percent less than the average. The moon is in apogee once every 27½ days. When perigee coincides with spring tides, the highest springs, called perigee springs, occur. These take place during a period of about three months in succession each year.

The earth varies in distance from the sun by about 1.7 percent greater or less than average. Thus, in January, the earth is nearest to the sun, or at perihelion. At this time the sun exerts its greatest tidal effect. Since the effect varies inversely as the cube of the distance, winter tides are generally 5 percent higher than average. In July the earth is most distant from the sun, or at aphelion, and its tidal effect is 5 percent less than average.

Still another fluctuation occurs as a result of the passage of the moon north or south of the equator, its declination. At its most northern declination, the moon's tidal effects will not be symmetrical; rather, one of the two semidiurnal tides will predominate in the Northern Hemisphere and the other semidiurnal tide will be strongest in the Southern Hemisphere.

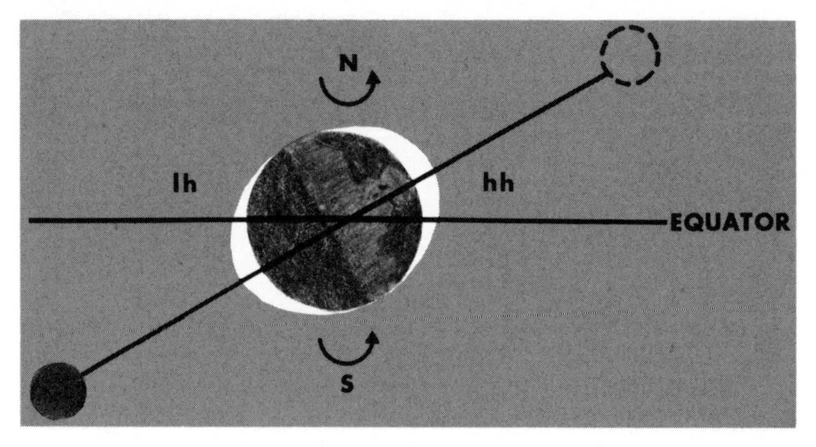

When the moon is in north or south declination (that is, north or south of the equator), the tidal bulge is no longer symmetrical. There will no longer be two equally high tides, but a high high tide (hh) and a low high tide (lh) each day. (Richard Marra)

The result of this will be that the two daily tides are unequal, with alternately larger and smaller high tides. A similar effect results from the sun's declination. In midwinter and midsummer, at the solstices, when the sun is farther from the equator, the effect will tend to cause unequal tides. Tides will tend to be greater at the equinoxes (mid-spring and mid-fall) and least at the solstices, when the sun is most distant from the equatorial plane.

The combination of astronomical positions that causes maximum tidal forces is when perigee spring tides coincide with perihelion and the moon and sun both have zero declination, with the moon, sun, and earth on a common axis. This rare event occurred in 3300 B.C., 1900 B.C., 250 B.C., A.D. 1433, and will happen again in A.D. 3800.

Summing up, then, the gravitational attraction of the sun and moon, combined with their movements relative to the earth and to each other, not only causes twice-daily tides, but also brings about considerable variations in tidal ranges at intervals related to the solar and lunar half-day, day, month, and longer periods up to thousands of years.

10. *Tidal Vagaries*

Unfortunately, the tides actually observed at various places on earth cannot be predicted from the theory of the gravitational effects of the sun and the moon alone. The various rhytms, of the lunar and solar day and half-day and the monthly variations and other periodic effects of the position of the sun and the moon, do exist, but the timing or phase and the effect of any one of these rhythms varies from place to place. In some places the half-day pulse is strong, with two equal tides a day, spaced 12 hours and 25 minutes apart. This is true for most Atlantic ports. In other places there is only one tide each day, spaced 24 hours and 50 minutes apart, as in some parts of the Pacific and the greater part of the Gulf of Mexico. In still other places, such as the Pacific coast of North America, there is a mixed tide, where one of the daily tides is higher than the other. This difference may be more pronounced at fortnightly intervals. In all these cases the tides are a whole or half lunar day apart. But at Tahiti high tide occurs at the same clock time each day, in a solar rather than a lunar rhythm.

According to Newton's theory, a bulge of water remains beneath the moon and travels westward as the earth rotates beneath it. But, in actual fact, high tide does not always occur when the moon is vertically in line with a place, at zenith or at nadir. There are varying amounts of delay, sometimes amounting to hours. This lag is known as the lunitidal interval. The average of this interval at new or full moon is referred to as the tidal establishment of the locality, differing from place to place. Again, the highest spring tide may not occur on the same day as a new or full moon, but may be as much as seven days late. This is known as the Age of the Tide. When the hottest part of a summer day comes not when the sun has reached its midday zenith, when one would expect

Moonlight has only a very small fraction of the strength of sunlight, but the gravitational effects of the moon are more than twice those of the sun. They are particularly evident in the diverse and complicated patterns of tides in the oceans and seas, along the coasts and in the estuaries. (Hedgecoth Photographers)

it to be warmest, but later in the afternoon, we see a similar phenomenon at work.

The Mediterranean is virtually tideless. There is a 6-inch spring tide range at Naples. On the other hand, tides range over 50 feet on the southern side of the Bay of Fundy and over 40 feet at Mont St. Michel off the coast of Normandy. And, in a few places, the tide that rushes up a river as a wall of water and is known as a tidal bore sometimes reaches 25 feet.

Why don't the tides behave the same in all parts of the ocean, strictly according to the lunar and solar rhythms? In the first place, Newton's theory of tides neglects the dynamic behavior of water in motion, which is affected by the depth and shape of the ocean. Secondly, the oceans do not cover the surface of the earth to a uniform depth but are divided into areas of varying size, shape, and depth, and are separated by land, which interrupts the free passage of the tidal wave.

The water in individual seas, bays, gulfs, and channels tends to form waves that have a natural period or speed, dependent upon the depth and shape of the water body. If the body of water is the right size, the direct action of the tidal forces may cause an independent tide. But tidal surges from

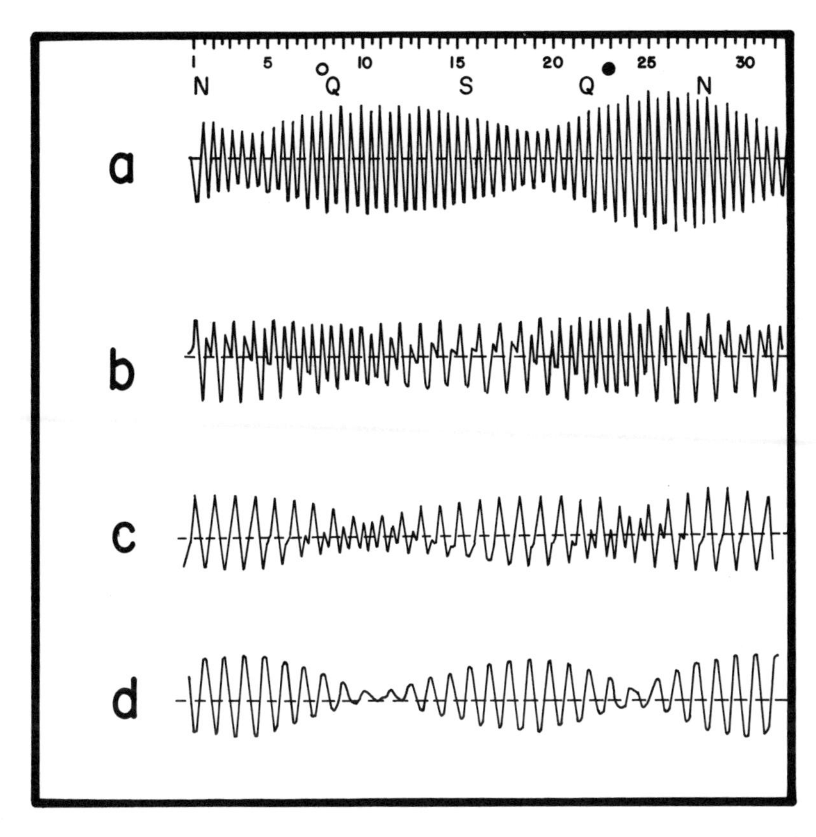

The principal kinds of tides are those that occur twice daily (a) and once daily (d). There are also mixed tides in which the daily or diurnal tide is predominant part of the time and the semidiurnal is predominant during the remainder of the month (b and c). The white and black circles under the dateline at the top refer to full and new moon. Q indicates the moon is over the equator. N and S indicate the times of maximum north and south declination of the moon. It will be seen in (a) that maximum tides (springs) appear just after new or full moon. In (d) maximum tides appear at maximum declination of the moon, when it is at its greatest distance from the equator. (a) Immingham, England. (b) San Francisco. (c) Manila. (d) Do San, Vietnam. (Richard Marra. From the Berlin Tide Table)

adjacent seas may also set the water in motion, causing sympathetic tides, which may not coincide with the independent tide. There are also inertial forces, resulting from the rotation of the earth, which set up tides, whose periods vary with the latitude (the half-pendulum day as explained in Chapter 11). The magnitude of these induced tides depends upon the degree of resonance, or coincidence of the period of the forced oscillation with the natural period of oscillation of the body of water.

The phenomenon of a natural wave period for any basin and the principle of resonance may be demonstrated by means of a simple pendulum. If a cord about 3 feet long, supporting a weight, is allowed to swing freely, it will immediately be seen that the time required for a complete swing to take place, the period of free oscillation, is one second. With a longer string the period would be greater, and with a shorter string it would be less than one second. In a similar way, if a large tub of water is tipped momentarily, a surge of water will go back and forth at a certain fixed speed and a period of oscillation measured in seconds. In a larger body of water, such as an estuary or enclosed sea, a free wave may be caused to oscillate, with a natural period of several hours.

To return to the pendulum, if the string is held in the hand and the hand moved steadily to and fro at one-second intervals, the pendulum will start to swing. The imposed or forced

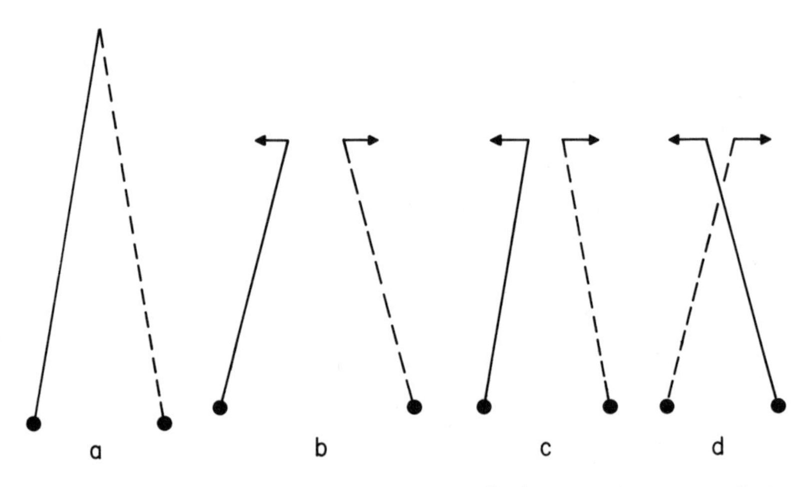

The pendulum demonstrates resonance, which is an important factor in determining the kind and magnitude of the tides. In (a) the pendulum is swinging freely at its natural period. In (b) the pendulum is swung by hand exactly in time with its natural period. The hand and the pendulum are in resonance, or in time. This results in a greater swing of the pendulum. In (c) the hand moves back and forth at a rate slower than that of the pendulum's natural period. The result is a motion in time with the hand but with a reduced swing or amplitude. In (d) the hand moves more quickly than the natural period of the pendulum. The pendulum moves, with reduced amplitude, but in the opposite direction to the hand. (Richard Marra)

movement of the hand has the same period as the free movement of the pendulum so that they are in step, or in phase. Under these conditions the movement of the hand continually reinforces the movement of the pendulum and it will increase the length of swing to a maximum. This reinforcement of movement when the force is in step with the free period is called resonance.

Next the hand can be moved at intervals somewhat greater than one second. The pendulum no longer swings at its natural one-second period but keeps time with the slower movement of the hand. However, in this case there is no resonance and the swings are much shorter. Finally, if the hand is moved more quickly with a period shorter than one second, the pendulum is again forced to swing at the same period as the hand but with a reduced swing. This time, however, the swing will be in a direction *opposite* to that of the hand.

This experiment demonstrates what actually happens when a tidal or other oscillation is applied to a sea or other basin of water. If the period of forced oscillation is the same as that for free oscillation in the basin, resonance occurs and higher tides take place in time with the pulse applied. If the period of free oscillation of the basin is less than that of the force applied, then reduced tides occur at the period of the applied force. If the period of free oscillation is greater than that of the applied force, a reduced tide occurs but lags behind by half a phase, or period, of oscillation.

It can now be seen why the concept of a bulge of water, or tidal wave crest, traveling around the earth as it rotates beneath the moon cannot explain the tides as we actually see them. One reason, of course, is that the ocean does not completely circle the earth. There are continents and land masses in the way. The other reason lies in the relation between long waves and the depth of water. The speed of a long or shallow-water wave traveling freely in the water is related to the depth of the water as explained earlier. The deeper the water, the longer and faster it is possible for a wave to become. The equilibrium tide of the moon in a canal circling the earth at the equator, and keeping pace with the moon above it, would have to travel at the speed of the earth's rotation, or once around the earth every 24 hours and 50 minutes, a rate of more than 1,000 miles per hour. At the equator a free wave could only travel at this speed if the water were 14 miles deep. In such depths the wave would be in resonance with the tidal forces and would reach enormous heights.

But nowhere on earth is the depth enough for a wave to be in resonance with the tidal forces. The free wave in actual depths must travel much more slowly than the tidal forces. Consequently, the forced wave would lag behind the moon by 6 hours, with the troughs beneath the moon instead of the crests.

At one time it was believed that the tide wave ran around the Southern Ocean, where there are no land obstructions. It was thought that this wave branched out into the Atlantic, Pacific, and Indian oceans and thence into the bays, gulfs and channels associated with them. Such a progressive wave does sometimes occur, as in Chesapeake Bay, where the tide moves as a wave from the mouth of the estuary toward the head, losing energy as it progresses.

It is now known that the tide does not usually function as a simple wave. Instead, each gulf, basin, or channel has its own characteristic oscillation, set in motion directly by either the tidal forces or by impulses that reach its entrance from the neighboring sea or ocean. In each case the depth and length of the basin determine the period of its free oscillation. According to its period it may be more or less in resonance with the half daily or daily tides. If it is in resonance, the tidal range will be large. If not, the range will be small. In addition, the effect of the earth's rotation in broad basins may be to develop a rotational tidal movement rather than simply a longitudinal movement.

The kind of tidal oscillation that usually occurs in a basin or sea is quite different from the progressive waves caused by the wind. Progressive waves continually advance in one direction. While the wave crest moves at a certain speed, the water particles themselves move in orbits, returning to the place from which they started. At the crest of the wave a water particle moves in the direction of the advancing crest. As the crest moves away the water particle moves downward.

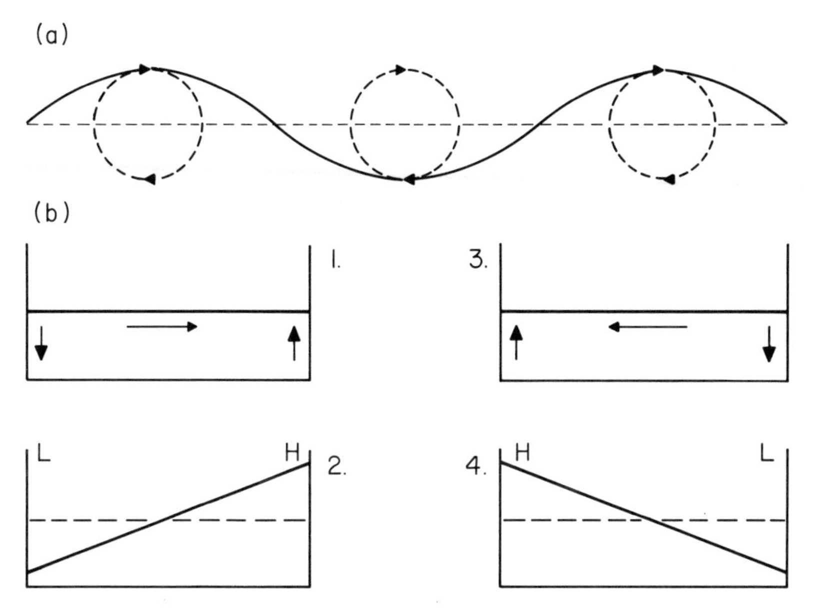

Waves may be progressive, like wind waves (a), or standing waves (b). In the first it will be seen that the water particles (dotted line) move in circular orbits and do not advance with the wave, moving from left to right. The water flows to the right at the crest (high tide) and to the left in the trough (low tide). In the standing wave, the water slops from side to side, but the wave does not progress. When it is high tide at one end of the basin, it is low at the other (2) or vice versa (4). The water does not move horizontally at the ends (antinodes) but only in between (1, 3) where the level is almost stationary. Greatest motion is halfway between high and low tides. (Richard Marra)

As the trough approaches, the particle moves in the opposite direction. Finally, before the next crest, it moves upward, completing a roughly circular orbit by the time it reaches the top of the crest. Thus, a progressive wave crest travels in a horizontal direction, but the water particles themselves do not advance.

Except in a few basins, such as the Chesapeake, tidal wave crests do not advance, but the water itself moves horizontally. This type of wave is a standing, or nonprogressive, wave. It is the kind that forms when a dish of water is tilted. The surface rises first at one end of the dish, then at the other end, while in the middle the water remains at a stationary level.

A standing wave is sometimes formed when progressive waves hit a seawall and are reflected as described in Chapter 5. The original waves and the reflected waves interfere with each other so that the observer sees waves that rise and fall in the same places without horizontal wave motion. As noted earlier, such waves are sometimes called clapotis.

In the case of a basin closed at both ends there can be no horizontal movement at the end, but only in the middle. The wave therefore oscillates about a stationary axis in the middle, which is called a node. At the node there occurs maximum horizontal flow at mid-tide and no change in level. At the ends there is no horizontal flow, but maximum vertical rise and fall. And, obviously, when it is high tide at one end it is low tide at the other.

When the water in a closed basin is set in motion by a disturbance, such as sudden changes in barometric pressure or squalls, the surface will continue to oscillate for some time with wave periods characteristic of that basin. The period of free oscillation is greater as the length of the basin increases and diminishes as the depth increases. It is thus possible to calculate the natural period of oscillation for any particular basin, and the calculated periods are found to coincide closely

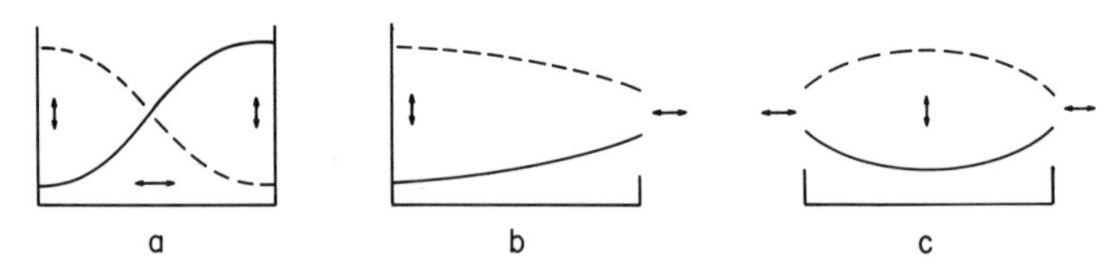

Tides in basins: In (a) there is no opening to the basin and a standing wave may develop in response to an external force (seiche), with a node in the middle. In (b), open at one end, and responding to a tidal oscillation from outside, a standing wave develops with a node at the opening, where maximum current flows, and an antinode at the closed end. If open at both ends (c) a node may develop at each end. (Richard Marra)

with actual observations. Non-tidal oscillations of this kind are known as seiches. Examples are those of Lake Geneva, where the calculated period between successive high levels is 74.4 seconds and actual measurements are 74.0 seconds, and Lake Vättern with a calculated period of 177.9 seconds and an observed period of 179.0 seconds.*

In a closed basin there can be no forced tide set in motion by tidal movements in neighboring seas. The only possible tide is the independent tide resulting from the direct action of the moon and the sun. Thus, in basins of limited extent, where the period of free oscillation is much less than tidal periods, there can be no resonance and the forced tidal oscillations will be very small in height. In large basins where the characteristic period is closer to the tidal rhythm, a greater tidal range will occur. If the period of free oscillation should coincide with the astronomical tide, then resonance would occur and much higher tides would result.

6. CALCULATED AND OBSERVED SEICHE PERIODS

Place	Period of Seiche in Minutes	
	Calculated	Observed
Lake Geneva	74.4	74.0
Lake Vättern	177.9	179.0

An example of a basin with a natural period that does not coincide with a tidal period is Lake Michigan with a characteristic period of 6 hours, where the tidal range at Chicago

* Standing waves in a closed basin may have more than one node, so long as the areas of maximum rise and fall are at the ends of the basin. With one node, high water at one end coincides with low water at the other, and the oscillation is one half of a full wave. With two nodes, high or low water occur simultaneously at both ends, so that the oscillation is a full wave. Thus, the wavelength of an oscillation in a closed basin is twice the length of the basin for a mononodal oscillation or equal to its length for a binodal oscillation ($\lambda = 2L$ where n is the number of nodes).

Since standing waves may be regarded as reflected progressive waves; since the period of a progressive wave in shallow water is $\lambda \div \sqrt{gh}$; and since the wavelength is equal to $2L$, then the period of a mononodal oscillation is $T = \lambda \div \sqrt{gh} = 2L \div \sqrt{gh}$. For an oscillation with n nodes this becomes $T = 2L \div \sqrt{gh} \cdot n$. Using units of hours for period, nautical miles for length of basins, and fathoms for depth, this now becomes approximately $T = 2L \div 8.2\sqrt{h} \cdot n$.

is 3 inches. In the western Mediterranean the characteristic period is 6 hours and the tide at Naples has a 6-hour range. The Baltic Sea has a characteristic period of 27 hours and a spring tide range of less than 2 inches at Marienleuchte.

Some seas or basins, such as the Irish Sea or the Strait of Magellan, are open to the sea at both ends. They oscillate with stationary waves, but in a different manner to closed basins, since the ends, being in communication with neighboring waters, allow free horizontal movement. Thus, there are nodes at both ends. The oscillation is a half wavelength. If only one end is open, then the free end becomes a node.

In the case of basins open at one end only, since tidal streams are able to flow at the open end only, this end becomes a node, the closed end an anti-node, and the simplest oscillation is a quarter wave. (The period for a free oscillation is therefore $T = \frac{4L}{n\mathrm{N}gh}$, according to the number of nodes.) Examples of such basins are the English Channel, the Bay of Fundy, Long Island Sound, and the Bristol Channel.

The seas into which the basins open have tidal oscillations, the effects of which, at the opening, may be sufficiently large to force a tide in the basin. If the natural period of the basin is close to that of the external tidal influence, then resonance will occur and the tidal range will increase.*

Table 7 gives the calculated lengths and depths of channels or gulfs that correspond to periods in resonance with the semidiurnal tide and can be expected, therefore, to respond by exceptionally large tidal ranges.

The dimensions of the English Channel, 25 fathoms (150 feet) deep on the average and 123 miles long, allow a free oscillation close to that of the semidiurnal period of 12 hours 25 minutes. Similarly, the Bay of Fundy, with a depth of 50 fathoms and a length of 175 miles, is resonant with the semidiurnal tide. In these places the tidal range is great.

---

* The dimensions of a channel that is in resonance with the semidiurnal tide are readily calculated from the expression $T = \frac{4L}{8.2\sqrt{H}\,(n)}$, by substituting for T the semidiurnal tidal period, or approximately 12.25 hours. The expression then becomes $L = 25n \cdot \sqrt{H}$ approximately.

7. DIMENSIONS GIVING RESONANCE WITH SEMIDIURNAL TIDE

Depth (fathoms)	Length (in miles)	
	Gulf, one end open $L = 25\ (n)\ \sqrt{H}$	Channel, both ends open $L = 25\ (n)\ \sqrt{H}$
12	87 (1)	175
25	123 (2)	247
50	175 (3)	349
200	349	698
1,000	780	1,561
2,000	1,104	2,207

(1) Long Island Sound; (2) English Channel; (3) Bay of Fundy.

It now becomes clearer why the tides come at different times at different places and why the tidal ranges vary so considerably. The tidal oscillations in basins and seas are mostly forced by tidal movements in the larger water masses and partly by the direct action of the moon. The tidal range depends upon the degree of resonance, or correspondence, between the natural period of the basin and the tidal periods.

11. *The Circular Tide*

One of the most important reasons for the varying nature of the tide in different places is the effect of the earth's rotation, the Coriolis force. This causes tidal currents other than very weak ones to be deflected to the right in the Northern Hemisphere and to the left in the Southern Hemisphere. The amount of deflection is proportional to the speed of the current, is greatest at the poles, and diminishes to zero at the equator, as explained in more detail in Chapter 15. Because of this, the simple wave movement of Newton's tidal theory may become a rotating movement, with the high tide crest moving in a circular path around the oceans and seas rather than in a straight line.

In a narrow channel, the banks are too close together for the rotary tide to develop. In such cases, though, the range of tides may become very much greater on one side of the channel than on the other, because of the development of what is known as a Kelvin wave. This effect takes place as a progressive tidal wave moves up the channel, as shown in the diagram. It will be remembered that in progressive waves the water particles at the crest of the wave move in the direction of the wave's path. In the trough the water particles flow in the reverse direction, as explained in Chapter 2.

As the high tide crest of the wave moves up the channel, the flow of water is deflected to the right, causing a higher tide on the right bank and a lower tide on the left. In the low tide trough, the flow of water reverses but is again deflected to the right of its path, so that there is now a higher low tide on the left bank and a lower low tide on the right. The net result is a greater range of tide on the right than on the left. This is the probable explanation for the situation in the western end of the English Channel, where the tides along

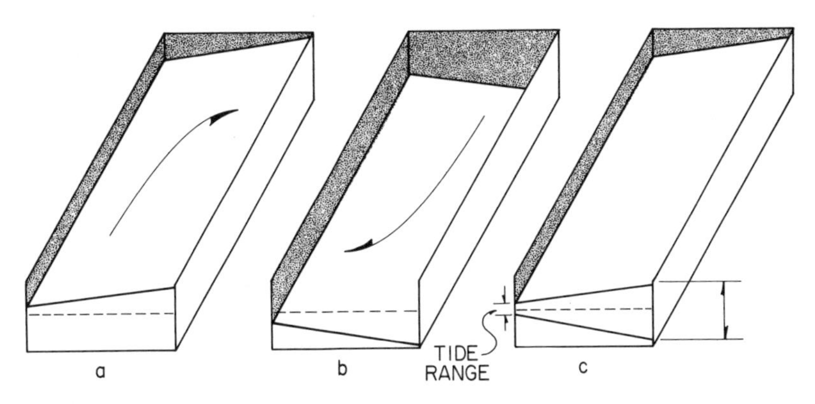

Narrow channels may develop a greater range of tides on one side than the other. This is due to the Coriolis force, which directs tidal currents toward the right in the Northern Hemisphere. As the tidal crest proceeds up the channel (a) the current moves toward the right, so that a higher tide occurs on the right bank than on the left. In the tidal trough, the receding current moving to the right (b) leaves a lower tide on the right bank than on the left. The result (c) is a greater tidal range on the right bank. Examples of this are the English Channel and the Bay of Fundy. (Richard Marra)

the southwest coast of England are smaller than those of the French coast. A related phenomenon exists in the Bay of Fundy.

In broader channels, the effect of the Coriolis force is to convert a standing wave into a rotary wave. This may be demonstrated by placing a cup or glass of water on a level surface and moving it with a circular motion. The water surface forms a crest that travels around the inner edge of the glass. On the opposite side of the cup to the crest is a trough, traveling in time with the crest. The surface at the center remains at a constant level.

The diagram shows how, in a square basin, the standing wave begins with a high tide crest at the far end. If there were no Coriolis force, the water would begin to flow from the far end to the near end, and at the end of 3 hours, a maximum water flow would occur at the center and a uniform level would be maintained throughout the basin. With the water continuing the flow, the tidal crest rises at the near end of the basin at the end of 6 hours. During the remaining 6 hours the process is reversed until at 12 hours high tide reappears at the far end.

The Coriolis force begins to distort the foregoing simple standing wave as the tide drops and water begins to flow. Since it turns to the right, at 3 hours a high tide will form to the left of the basin and low tide to the right. The water will then begin to flow from the new high tide to low tide

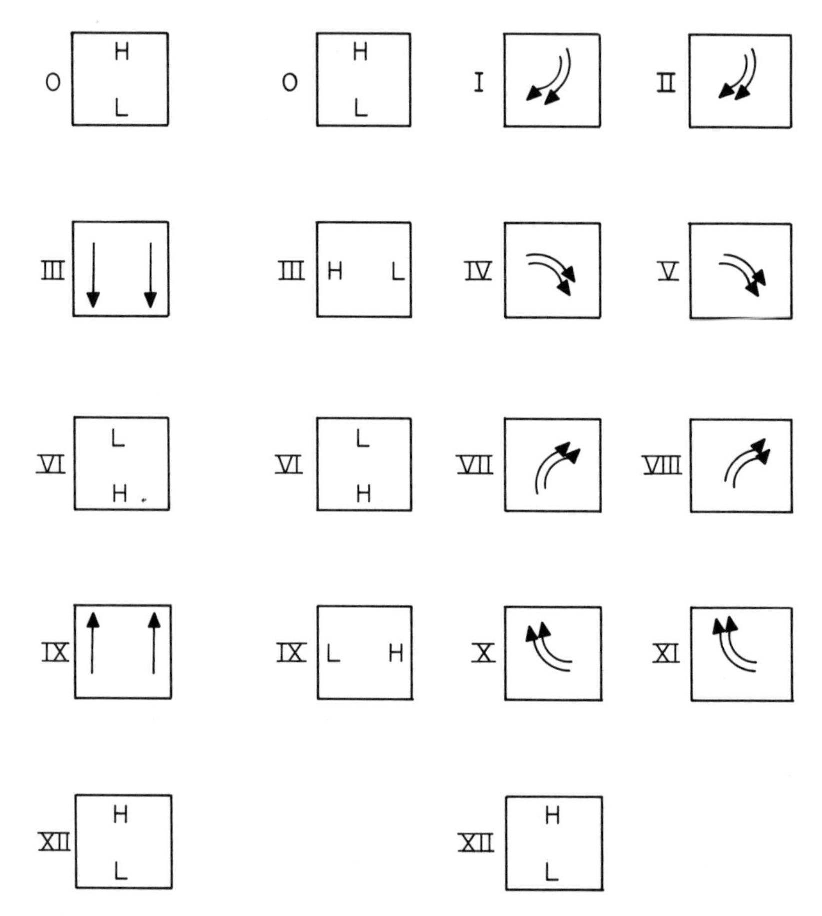

In broad channels the Coriolis force causes a rotary tide. The left-hand column of diagrams shows the progression of a standing tidal wave without the Coriolis force. The Roman numerals show the passage of time in hours, starting with O, when high tide is at the top of the diagram. The tide simply moves back and forth from top to bottom. The other diagrams show the tide under the influence of the Coriolis force. As the tide runs toward the bottom of the diagram (I and II) it is deflected to the right, resulting in a high tide toward the left (III). The tide now runs from high at the left to the low level at the right (III), but is again deflected to the right (IV, V), resulting in a new high water at the bottom of the diagram (VI). The net result is that high tide moves in a counterclockwise direction around the basin. (Richard Marra)

level, again turning to the right, until at 6 hours the crest is formed at the nearer end of the basin. During the remaining 6 hours the crest continues its path around the basin, with the trough following it on the opposite side of the basin.

The result is a high water crest moving in a counterclockwise direction around the basin. The currents flow in the direction of the crest at high tide and away from it in the trough at low tide, in the manner of a progressive wave traveling in a circle. Instead of oscillating back and forth across a stationary nodal line, as in a stationary wave, the tide revolves in a circle around a central nodal point, where the level is practically stationary. The tidal current simply alternates in a to and fro movement.

Upon reflection it will be seen that the amphidromic oscillation described could be arrived at equally well as the result of interference between two standing waves, one of which runs north–south and the other east–west, each with a 12-hour period and equal amplitude, but with a 3-hour, or quarter wave, phase difference.

When a Kelvin wave is reflected by the end boundary of the channel, it interferes with the incoming wave, and its asymmetrical nature gives the resulting standing wave a rotary motion, in which the wave crest instead of oscillating around a nodal axis, will rotate around a nodal point. This amphidromic wave is similar to the amphidromic wave that occurs as a result of the action of gyroscopic forces upon a standing wave, but the streams, or orbital movements of water particles are different.

In a small, deep sea, where tidal amplitude is small and currents are weak, the effects of the Coriolis force are negligible. Under these circumstances the direct tractive forces of the moon become apparent. The result is a rotating tide of small range, in which the cotidal lines progress in a clockwise direction. Typical examples of this are the Caspian Sea and the Black Sea. It will be noted that rotation is in the opposite direction to that caused when a standing wave responds to the Coriolis force.

Another type of amphidromic wave may be caused by interference between longitudinal and transverse oscillations in a basin. Under these circumstances, depending upon the phase difference, the movement of cotidal lines may be either clockwise or counterclockwise. Since most amphidromic tides are counterclockwise, it is therefore probable that they result from gyroscopic effects rather than interference.

The gyroscopic effect of the earth's rotation upon waves is

only significant for waves with periods of the order of magnitude of the inertial oscillation or half pendulum day. In most places this is true of either the diurnal or semidiurnal tides.

The period of the half pendulum day is always greater than 12 hours. If a pendulum is allowed to swing freely back and forth it will continue to do so in the same plane while the earth rotates beneath it. Thus, at the North Pole, the pendulum, swinging across, for example, the 0-degree and 360-degree earth meridians, would continue to swing in the same celestial plane while the earth's meridians rotated beneath its plane of oscillation. Twelve sidereal hours later, the 360-degree and 0-degree meridians would once more appear beneath the plane of oscillation of the pendulum. This period is the half pendulum day. It is equal to 12 divided by the sine of the latitude. At the equator, where the horizontal component of the earth's rotation is zero, the plane of the pendulum swing would not appear to rotate at all. The pendulum half-day would be infinite in time. At 30 degrees latitude it would be 24 hours, at 60 degrees latitude 14 hours.

Thus, in most latitudes, in the absence of friction forces, the inertial effects of the earth's rotation have a period of the same order of magnitude as the movements of rotating tides, with periods ranging from 12 hours to more than 24 hours, depending upon latitude.

8. VARIATION OF PERIOD OF HALF-PENDULUM DAY WITH LATITUDE

Latitude	Period of Half Pendulum Day
90°	12 hours
60°	14 hours (about)
30°	24 hours
0°	∞

These rotary tidal systems, known as amphidromic tides, are very common in the oceans and seas, where they are set in motion by tidal waves in adjacent seas, according to the prin-

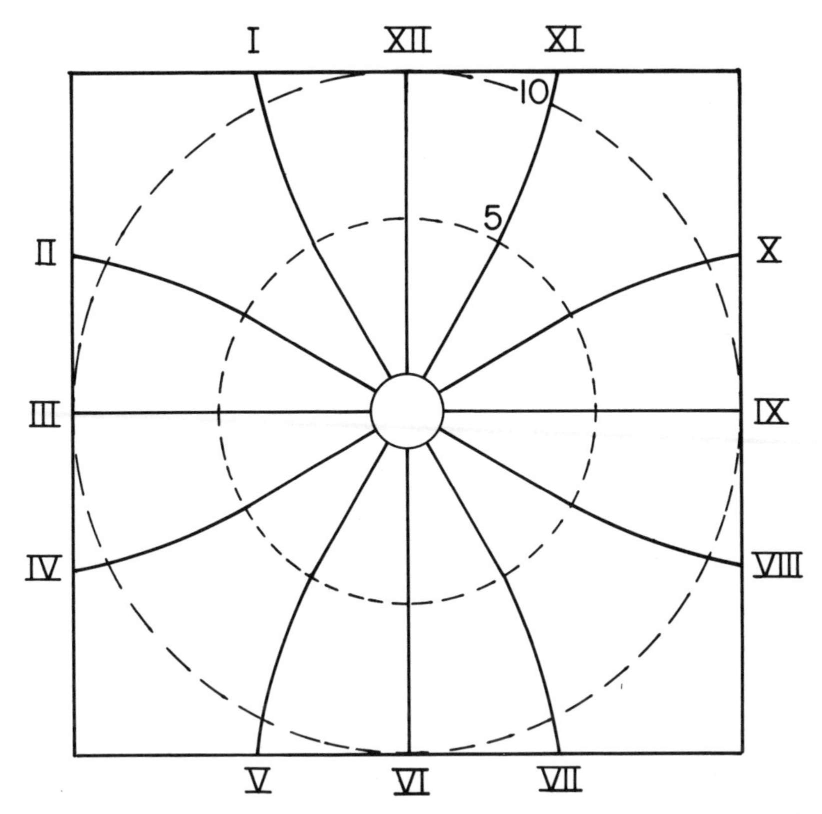

Tidal movement in a broad basin or sea is rotating. At the center of the basin there is little or no tidal movement. This is the nodal point. Radiating from it are lines joining places where high tide occurs at the same time (shown in Roman numerals). Thus all points on the line marked XII have high tide at twelve hours. One hour later high tide occurs on the line marked I, and so on with the tide moving in a counterclockwise direction. The lines are co-tidal lines. The dotted lines join places with similar tidal range, 0 feet at the nodal point, 10 feet at the circumference, and 5 feet at an intermediate place. These are co-range lines. The hours shown are the time lapse between the moon's transit, either at Greenwich or a specified place, and the appearance of high tide. (Richard Marra)

ciples of resonance explained in Chapter 10. Because of their importance, a simple means of showing them on a chart is described.

At the center of a rotary tide is the stationary nodal point. Radiating from this are cotidal lines, which join all points at which high tide occurs simultaneously. Thus, at all points on the co-tidal line marked I, high tide will occur one hour after the moon's transit. One hour later high tide will have reached the cotidal line marked II, and so on, progressing around the circle. Charts frequently show other lines at right angles to the co-tidal lines that form circles with the node as center. These join points with equal tidal range. At the node the range is zero. Extending outward from the node, these co-range lines show increasing tidal range.

The tidal currents associated with a rotary tidal system also follow curved paths. Near the node the currents follow an almost circular path, but toward the edge of the basin they flatten out and become linear to and fro movements.

With an understanding of resonance, standing waves, and amphidromic tides, it now becomes much easier to understand why the actual tides vary so much from place to place.

The resonance of a basin may cause it to respond more to one tidal rhythm than another. The tidal rhythm most in evidence is usually either the semidiurnal or diurnal tide. But in some cases the solar tides may become important. At some locations, the fortnightly spring and neap tides, because of the relative positions of the sun and the moon, exert maximum effect, as in the waters around the British Isles. In other places the declination tides, because of the movement of the moon in its orbit north or south of the equator, may be especially marked, as in parts of Borneo. In the Bay of Fundy the perigee–apogee tides, depending upon the distance separating the moon and the earth, are important.

The tides in a bay or sea also depend upon the nature of the tides in the neighboring seas with which they co-oscillate. There may also be various combinations of progressive, standing, and rotary waves, together with the modifying effects of friction and reflection.

In the South Atlantic (see maps) the tide appears to be predominantly a progressive semidiurnal wave, set in oscillation by the tidal wave of the Antarctic Ocean. This progressive wave moves from south to north over the entire width of the ocean, taking 6 hours to travel from the tip of South Africa to the equator and 6 more to reach the latitude of the Caribbean. A nodal point appears to exist in the eastern Caribbean, since the tidal range at Puerto Rico is less than 1½ inches. North of this point the North Atlantic oscillates around a node to the south of Greenland.

The tides of the Pacific Ocean are less well known than those of the Atlantic. In general they are semidiurnal, except

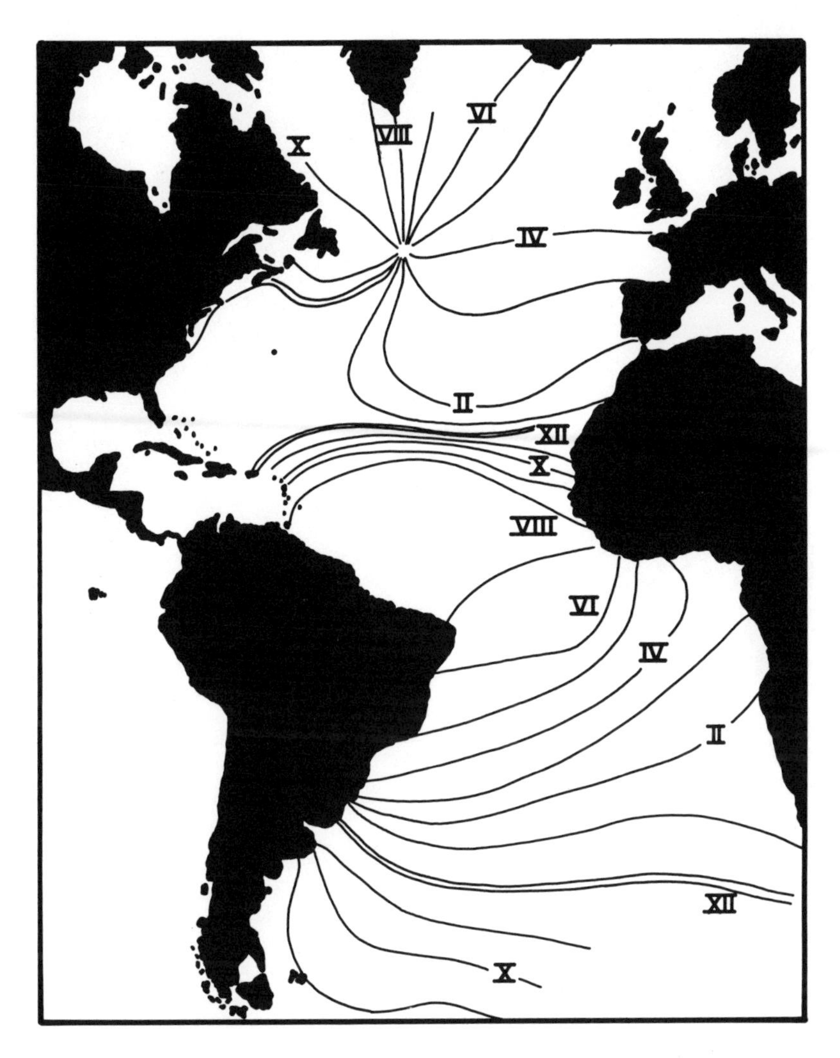

The Atlantic tide is mainly semidiurnal with a marked node in the North Atlantic and a weak node near Puerto Rico. The tide starts in the South Atlantic (XII) and sweeps northward (II, IV, VI), until, twelve hours later (XII), it has passed the equator. In the North Atlantic the tide moves around the ocean, northward along the coasts of Africa and Europe (II, IV) toward Greenland (VIII) and then southward down the coast of North America. (Richard Marra, after Hansen)

in the Gulf of Alaska from Seattle to the eastern Aleutians, the Antarctic, New Guinea, Formosa, Manila, Hawaii, and the Solomons, where the diurnal tide prevails. Along most of the Pacific continental coasts the diurnal tide, though not predominant, is stronger than in the Atlantic. In the vicinity of the Solomon Islands and to the south of the Gulf of Alaska are nodal points, corresponding to small tidal ranges. Two other nodal points are believed to exist southwest of the Gulf of

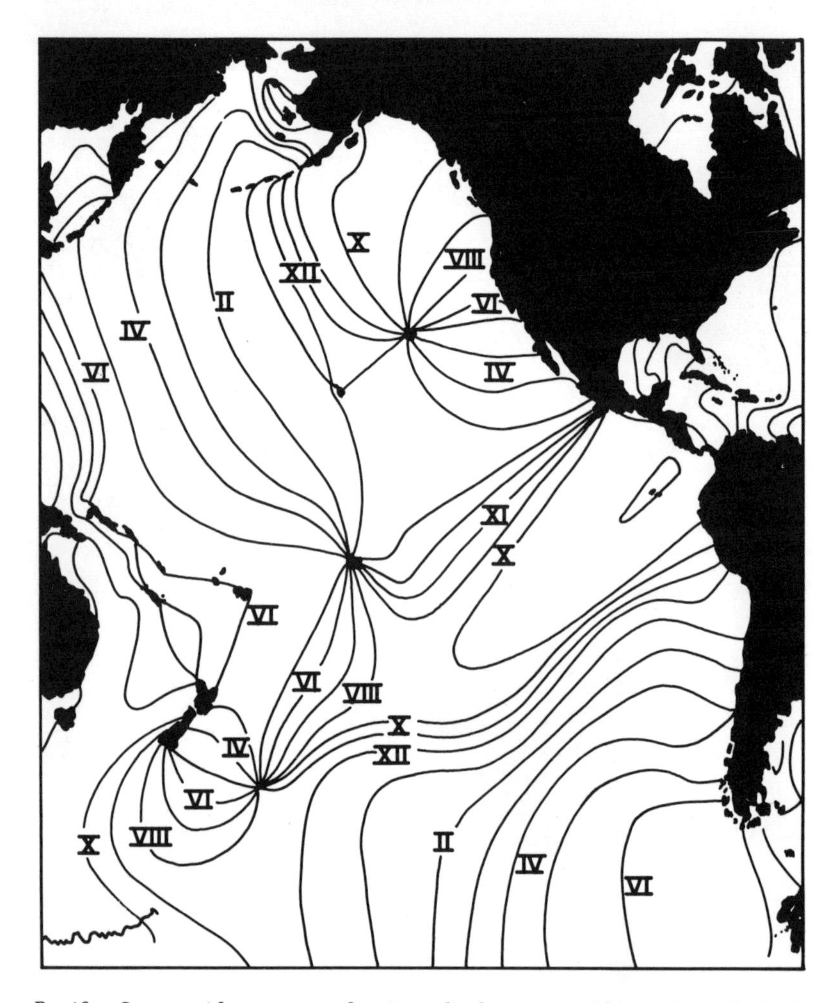

Pacific Ocean tides are predominantly lunar semidiurnal, with high tide every twelve hours. The co-tidal lines show that the tide in the northeast Pacific rotates around a nodal point off the coast of California. The tidal rise first begins at the coast of Baja California (IV). Two hours later it has progressed up the coast (VI). Six hours later it has reached Alaska (X, XI). Also shown is a nodal point at Tahiti, where there is no lunar tide, but a solar tide only. (Richard Marra, after R. A. Harris)

Panama and to the south of Easter Island. Along the west coast of North America, as the map shows, high tide first occurs at San Diego. Three hours later it has advanced to Sitka, Alaska, and at the end of another 3 hours it has reached Dutch Harbor, Alaska.

Tahiti has tides of small range, but they occur at the same time each day, instead of 50 minutes later. In fact, the native word for midnight is also the word for high tide. The reason for this is that Tahiti is located near a nodal point of the lunar tide, and therefore does not respond to the lunar rhythm. In the absence of the lunar tide, the small solar tide, which would otherwise be obscured, is able to become noticeable. Clearly, the nodal point of the solar tide must be located at some distance from the area.

In the Bay of Fundy, which has already been mentioned, it is not unusual to see a ship that is ready to steam away from a wharf still high and dry in the mud. The crew knows that the vessel will soon be floating in deep water. At Minudie on Chignecto Bay, Nova Scotia, fishermen string nets on 15-foot poles stuck in the sea floor across an inlet when the tide is out. The incoming tide not only submerges the nets, it also brings in schools of fish. When the tide recedes, the fishermen drive horse-drawn carts over the dry sea floor and, ascending ladders, pluck fish from the nets.

The tidal range of the Bay of Fundy, the greatest on earth, clearly results from resonance with the diurnal tide of the ocean, which reaches its mouth and stimulates a standing wave, with the node at the entrance. Its natural period of resonance is about 6½ hours, which is almost exactly in tune with the semidiurnal tide period. This strong resonance is responsible for the huge tides. Being a standing wave, the lowest tides are at the node, or entrance, and greatest toward the head of the bay. Thus, at Nantucket Island off the coast of Massachusetts the mean range is 1.2 feet, but it is 20.9 feet at St. John, Canada, and 44.2 feet at Noel Bay. A contributing reason for the greater tides at the head of the bay is the funneling effect of the narrowing width and shallowing of the channel. Also characteristic of a standing wave is that the high tide occurs at almost the same time at the mouth as at the head, being only 24 minutes apart over a length of 162 miles. The highest tides are the perigee spring tides. At perigee, tides reach 50 feet, at apogee only 40 feet.

As explained earlier, a bay such as this is too narrow for the Coriolis force to develop a rotary tide. Nevertheless, there is a cross-channel effect of the earth's rotation whereby the tides are greater on the south shore than the north shore. On the north shore the tide ranges extend from 1.2 feet at Nan-

9. THE MEAN PROGRESSIVE TIDAL RANGES IN THE BAY OF FUNDY

North Shore		South Shore	
	feet		feet
Nantucket Island	1.2	Cape Sable	9.0
Minomy Point	3.7	Yarmouth	14.0
Nanset Harbour	6.0	Grand Passage	18.2
Gloucester	8.9	Digby Gut	24.1
West Quoddy Head	15.7	Port George	27.8
St. John	20.9	Black Rock Light	31.5
Quaco	26.3	Horton Bluff	42.0
Folly Point	39.4	Noel Bay	44.2

tucket to 39.6 feet at Folly Point. On the south shore the ranges are greater, from 9.0 feet at Cape Sable Island off the coast of Nova Scotia to 44.2 feet at Noel Bay. Another spectacular feature of the bay is its tidal bore, which will be described in Chapter 13.

The weak tides of the Gulf of Mexico do not exceed 2½ feet. Because of its dimensions, the Gulf is only in weak resonance to the tidal rhythms, responding to the diurnal tides, but scarcely at all to the semidiurnal tide. As a result the Gulf tides are diurnal everywhere except on the west coast of Florida. The tide appears to oscillate with the tides of the Straits of Florida, which sweep around the Gulf and leave by the Yucatán Channel. Tidal variations are governed by the extremes of lunar declination rather than by the phase of the moon. A similar situation exists in the Caribbean.

It has already been mentioned that in the English Channel a progressive Kelvin wave moves from the west up to the region of Ventnor, with high tide occurring at Ventnor. From the Isle of Wight to Dover, high tide occurs everywhere within about 10 minutes, as will be seen on the map (page 146). This is characteristic of a standing wave. The convergence of the co-tidal lines toward a point west of the Isle of Wight and the

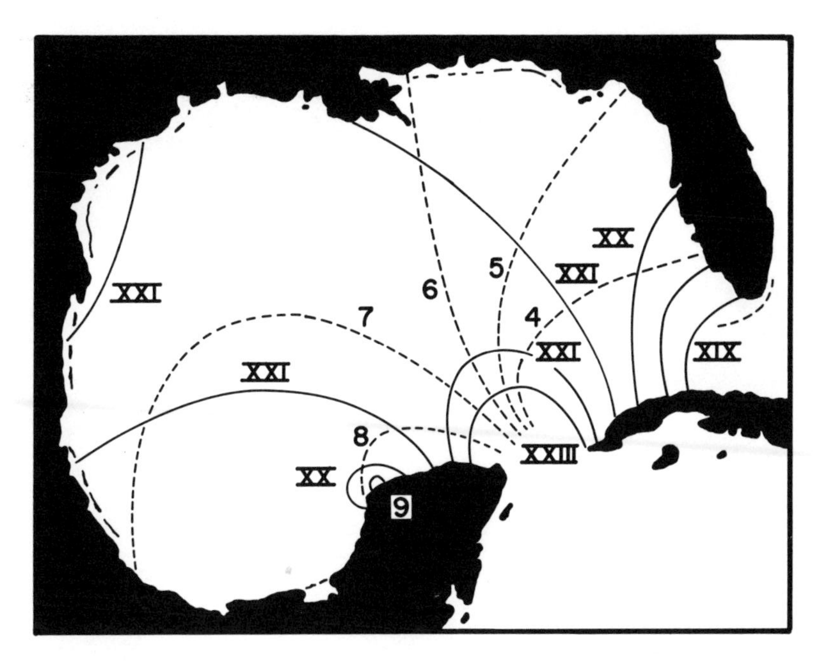

Tides in the Gulf of Mexico are weak. The co-range lines (broken lines) show ranges of 4 to 8 inches. The tide progresses, as shown by the co-tidal lines (solid lines) from the Straits of Florida (XIX) up the west coast of Florida (XX) and down the coast of Texas and Mexico (XXI). (Richard Marra, after Grace)

reduced height of the tide in that area suggest a weak tendency toward a rotary tide, which cannot develop properly in the narrow Channel.

The North Sea has greater tides on the west than the east, and the timing from place to place at first appears to be very complicated. It is generally agreed that the North Sea oscillates in response to the North Atlantic tides entering from the north. Since the opening to the English Channel is relatively small, the sea acts as a broad bay open only to the north. Apparently the semidiurnal wave, which is completely dominant, moves southward as a progressive wave until it is reflected at the southern boundary. Because of the Coriolis force it becomes a Kelvin wave. Interference between the original wave and the reflected wave from the south sets up a double rotary system. The main nodal point is off the coast of Denmark, a weaker one is to the south, and another small point occurs off the southern coast of Norway near Stavanger. The reason for the displacement of the two northerly nodal points to the east appears to be the frictional effect of the shallow banks of the North Sea. The result is that greater tidal ranges occur on the coasts of England and Scotland than on the eastern shores

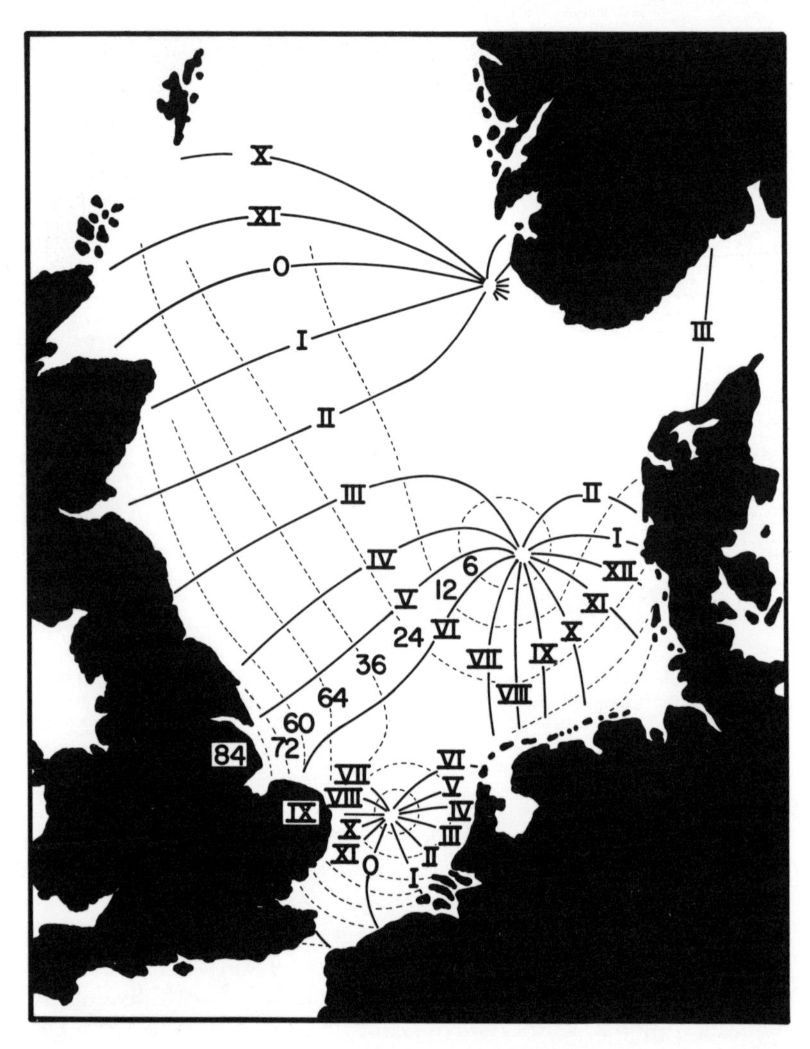

The North Sea has three nodal points around which rotary tides swing. The tide arrives from the North Atlantic (X) and swings around a weak nodal point off Sweden. In the eastern part of the North Sea is another nodal point. The third rotary system is in the extreme south. The broken lines and the Arabic figures show the co-range lines and tidal range in inches. The range is much greater on the English coast (72 inches) than on the continental coast, the result partly of frictional effects. (Richard Marra, after Proudman and Doodson)

of the North Sea. The diagram shows how the high tide progresses at various rates from place to place along the coasts.

The Black Sea and the Caspian Sea have tides that at first do not seem to follow the usual pattern. The tides in these seas are rotary tides of very small range, which move clockwise instead of in the counterclockwise direction that the

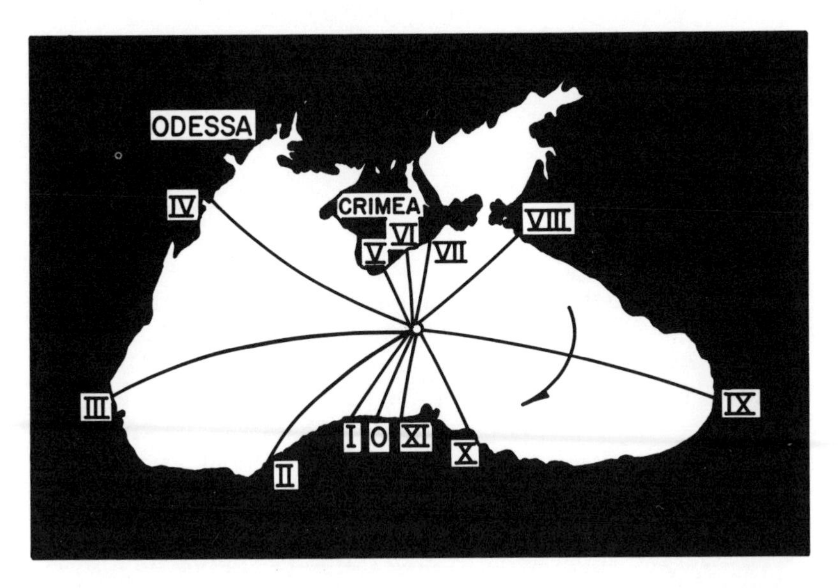

Unusual tides in the Black Sea move in a clockwise direction starting from the south (O) and appearing at Odessa 4 hours later (IV), then completing the circle by the Crimea (VI) down the eastern coast of the sea. This rotary tide is not due to the Coriolis force but to the direct effect of the moon's movement. (Richard Marra, after Macmillan)

Coriolis force induces. The explanation of this peculiarity is that the tides in these small, deep seas result directly from the gravitational effect of the moon, which causes the tidal wave to follow the moon clockwise, as the earth rotates. The tidal streams are very slight, so that the Coriolis force, which depends on the strength of the current, is negligible and fails to develop the counterclockwise course usually characteristic of rotary tides.

Another peculiar tide occurs at Southampton, England. The port is connected with the sea by two arms of a channel, the Solent, each arm passing on a side of the Isle of Wight. The tides of Southampton remain for a long period at high water, and at such phases there may even be a double high water between the normal low tides. Under these conditions high tide occurs early, drops slightly for a while, and then returns to high tide. The probable cause is the existence of different levels in the two arms of the Solent, which set up hydraulic gradients.

In a number of estuaries the period of flood tide is much shorter than the period of ebb. This occurs because progressive tide waves, just as in the case of wind waves, become steeper in front and longer behind the crest when they enter shallower water, as explained in Chapter 5.

Thus, although tidal behavior varies considerably from place to place, its departure from the simple tidal theory of Newton can be explained in terms of the dynamic behavior of water in motion, the effects of resonance, and the Coriolis force. But these modern concepts still do not enable us to predict tides with accuracy. For the practical guidance of fishermen, yachtsmen, and even shell collectors, not to mention those who design large tidal energy and power plants, it is necessary to adopt an altogether different approach.

12. The Real Tide

From the earliest times fishermen and sailors have noted a relationship between the moon's phases and the state of the tides. This was the beginning of the modern system of tidal predictions, which is based upon observations. These observations, initially, related the times of high tides compared to the meridian passage of the moon. They also noted the increase and decrease of the tidal heights in relation to the phase of the moon. From these first observations, simple guides and tables were prepared to inform seafaring men at the major ports.

One of the first crude tide tables is contained in a manuscript now in the British Museum. This tide table gives the time of floodwater at London Bridge in relation to the phase of the moon or the number of days after new moon. It was drawn up by the abbot of St. Albans in 1213. Another partial aid to predicting tides appeared in the Catalan Atlas prepared for King Charles V in 1375, which gave data on the local establishment of the tide for a number of places. But the precise tabulation of predicted times and heights for every day of the year was not achieved until more than four and a half centuries later.

Newton clearly established the relation between the gravitational forces of the sun and moon and the tides on earth, but his equilibrium theory could not explain the tidal movements as they actually existed. However, the French scientist P. S. Laplace (1749–1827) separated the various components of the astronomic tides, breaking them down into simpler components that represented the effects of the sun alone, the moon alone, and the individual effects of declination, distances between moon and earth, sun and earth, and other causes of fluctuation. This is essentially similar to the spectral analysis of wind waves mentioned in Chapter 4.

Lord Kelvin was responsible for systematizing Laplace's

The reversing falls created at the mouth of the St. John River as it enters the Bay of Fundy through a narrowing gap. At certain stages of the tide there is a turbulent rush of water toward the bay. As the tide rises, the incoming stream brings the water on both sides of the gap to the same level, but as the incoming tide continues to rise, the downstream level increases and the falls then run in the opposite direction, up the river. (New Brunswick Travel Bureau)

Huge tide changes are recorded by a gauge in the wooden casing at the top of these piles. The tidal difference shown here, more than 30 feet, is being recorded at Anchorage, Alaska. (U.S. Coast & Geodetic Survey)

method by analyzing curves of the actual observed tide at any one place into tidal curves, or harmonic components, corresponding to each of the astonomic tidal constituents. Each of the astronomic forces is supposed to produce a wave whose period, or time from crest to crest, is the same everywhere. But the phase, or the time lag behind the transit of the moon differs from place to place on earth and so does the amplitude or tidal range. These must be measured. If all of the tidal components at one place are then added together, with the appropriate amplitudes and phases, along with the predicted future positions of the sun, moon, and earth, the result is the actual future tide. In order to predict tides it is, therefore, first of all necessary to measure the tide at given places over a period of time covering at least one year and ideally nineteen years.

The tides are measured by means of special gauges that

record the height of sea level continuously but ignore the short period movements of wave and swell. Basically these gauges consist of a float on the surface of the water that is connected by a wire running over a pulley to a counterweight. As the water rises and falls, the wire moves a pen back and forth over the surface of recording paper on a drum rotated by clockwork, and so records the curve of the rise and fall of tide. The float is usually enclosed in a well connected with the sea by a narrow pipe. Because of the restrictive effect of pipe, the more rapid movements of sea level resulting from wave and swell do not disturb the level in the well, so that the gauge only records the slower tidal movements.

Other methods of measuring tides have also been introduced. One of them is essentially an instrument for measuring pressure changes. When placed on the sea floor the instrument responds to the changes in pressure resulting from the fluctuating height of the water column above it. Such instruments are able to record wave heights as well as changes in depth resulting from tidal movement. They are especially useful at distances from the shore, where it is impractical to erect a structure to house the regular type of gauge. The gauges may be electrically connected to the shore for purposes of recording. Where a continuous tide record is not required, a pole, graduated in feet and inches, may be driven into the bottom, or attached to a permanent dock structure. Visual observation makes it possible to determine such data as the lag of the tide compared to the moon's transit or to a reference station, and also the range of tide.

The tidal record is broken down into the various tidal constituents. Each of these is averaged out for a long period of time and their phases and amplitudes are determined. When this has been done a number of irregular water movements still remain. These are due to storm surges and other local disturbances not connected with the sun and moon. These sometimes disastrous freak tides must be computed separately, where possible, usually only a few hours in advance.

The work of analyzing tidal records is carried out by dividing the observed tidal curve for a year or more into the period

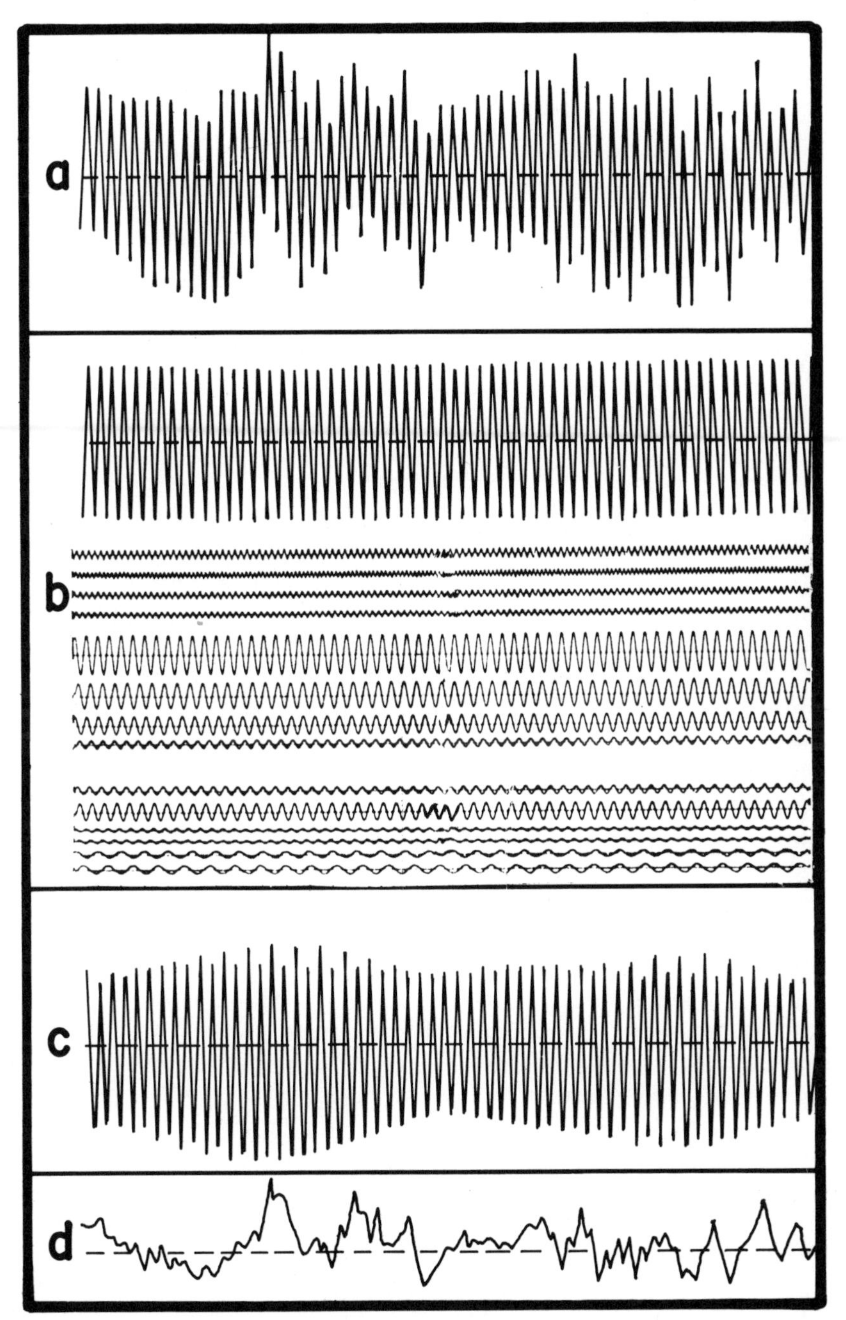

Analysis of tidal movements: The top curve (a) shows the changes in sea level recorded by a tide gauge. The 15 curves below this (b) show the different harmonic constituents, which have been derived from (a). When put together again they give the curve shown at (c). This shows the tide that would have occurred if there were no meteorological effects. Curve (d) shows the meteorological effects that still remain as part of the actual curve (a). (Richard Marra)

of each constituent tide in turn. For the solar tide, the period would be 12 hours and this in turn broken down into one-hour units. For each of these hours the amplitude is measured. Then the amplitudes for this particular hour for every day of the record are added together and averaged out. When each hour of the day has been so treated and the averages are put together, what is left is the tidal curve for the particular constituent. All of the others have been averaged out. Each constituent in turn is treated in the same way. For the principal solar constituent alone, averaging over a nineteen-year record would involve the considerable total of 166,440 measurements!

The phases and amplitudes of each constituent always remain the same at a particular place. The principal constituents are the main lunar, M_2, with a period of 12.42 solar hours; the main solar, S_2, with a period of 12 solar hours; the lunar constituent resulting from variation in the moon's distance, N_2, with a period of 12.66 solar hours; and the solilunar constituent resulting from changes in declination of sun and moon throughout their orbital cycle, K_2, with a period of 11.97 solar hours. All of these have two cycles each day, designated by the subscript $_2$. There are also diurnal constituents and the moon's fortnightly constituent. These eight components are generally sufficient for accurate prediction, but the more components that are used, the more accurate the predictions will be. Once the relative positions of the moon, sun, and earth are calculated, the tide can be predicted at any future time by using the known amplitudes and phases of each constituent.

Once analysis of the tide data has been completed, the information is placed into a tidal prediction machine. The first of these was built by Kelvin in 1882. Later, a machine was built in Germany that could use twenty tidal constituents. This machine took nearly 15 hours to calculate the annual tide tables for a single harbor. One of the latest machines of this kind, in Hamburg, uses as many as sixty-two constituents.

The principle of the tide-predicting machine is very simple. Its purpose is to mechanically produce the constituent oscillations with correct amplitude and phase and then add

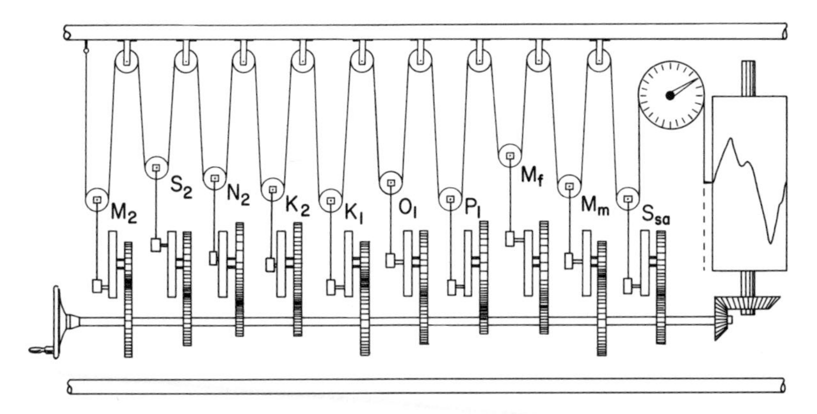

Tide predictions are made by means of a machine similar to that illustrated below. The constants for the various harmonic constituents, obtained by analyzing the actual tidal records (see previous figure), are put together in this machine, which then draws the curve of tidal predictions. The diagram above shows the working parts. The M_2 constituent, for instance, determines the gear ratio, and thus the period of oscillation, for the gear on the left-hand side. The placement of the crank, which it drives, determines the length of movement and thus the amplitude. Both period and amplitude are thereby transferred to the pulley above and this in turn moves the cable around the pulley. A similar transformation is made with the next gear, crank, and pulley for the S_2 constituent and so on. When the handle at the left is turned, it actuates all of the components and their combined effect is transmitted by the cable to the pen at the right, which records the predicted tides on the recording paper on the drum. Below is the tide-predicting machine at the Liverpool Tidal Institute, England, which is able to incorporate forty-two constituents in its predictions. (Liverpool Tidal Institute)

them together to give the predicted tide curve. This is demonstrated in the diagram. Each moving pulley represents a constituent. It is moved up and down by a special crank, so constructed that the distance of up and down movement corresponds to the tidal amplitude. The rate of rotation of the crank, and therefore the period of up and down motion of the pulley, corresponds to the period of the constituent. A pen attached to the end of a cord running over the pulley would draw the tide curve for the particular constituent. In fact it is similar to the tide gauge described earlier, except that the crank provides the motion instead of a float and it only draws the curve for a single constituent tide. But the machine has as many pulleys as constituents, all arranged in their proper phase and with the correct amplitude and period, and with the same cord passing over each pulley in turn. As a result, the pushing and pulling of all the pulleys at once combine to make the final motion of the pen at the end, which is the reconstructed tide.

In recent years, the mechanical tide prediction machine has begun to be replaced by modern computer methods, which considerably reduce the time involved. The calculations are made by feeding into the machine a tape containing the astronomical variables for the year together with another tape that provides the measured phase and amplitude of the tidal constituents for the port in question. The results are automatically typewritten. Compared to the 15 hours required by the mechanical prediction machines, the computer can produce the typewritten table for one port in less than one hour. The accuracy of most of the machines in use is close to one minute of time and one-tenth of a foot in tidal height.

Since the methods used in tidal prediction are based upon measurements of the actual tides, they do nothing to add to our knowledge of how tides come about. Nevertheless, by the method of harmonic analysis, very accurate predictions of tides can be made. It is even possible to apply this method to predict to some extent abnormal situations, such as the double high tides that exist in the Solent as a result of gradient currents. In this case, the curve for M_2, the principal semidiurnal tide, when combined with the M_4, or quarter

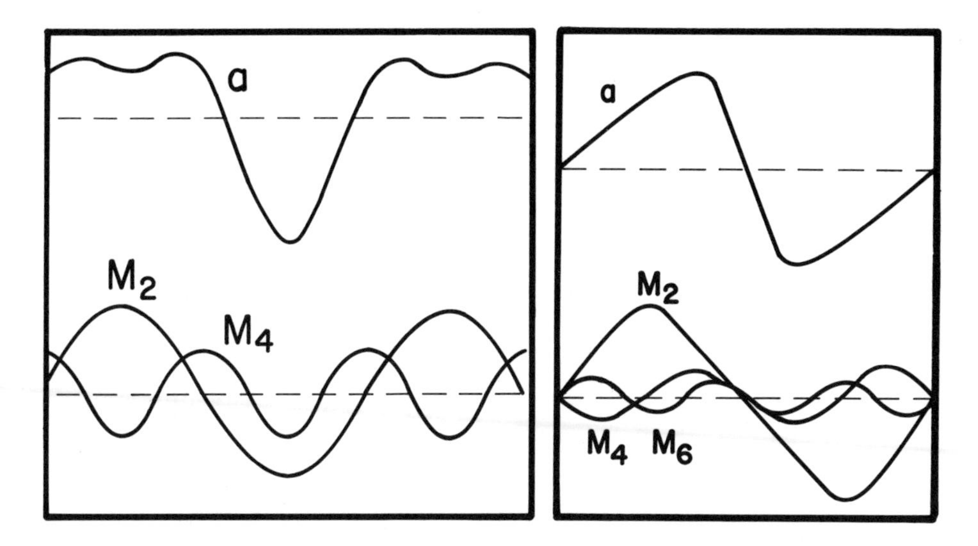

Unusual tides: The figure on the left illustrates the way in which a double high-water tide (a) may be predicted by a suitable combination of the M_2 and M_4 constituents obtained by analysis of the observed tide. To the right, a typical shallow-water tide (a), with a shorter flood period and a longer ebb, is predictable by adding together the M_2, M_4 and M_6 constituents, derived from the observed tide. (Richard Marra)

diurnal tide, using the constants of phase and amplitude derived from analysis of tidal measurements, gives a reasonably good prediction of this peculiar tide, as the diagram shows. Again, it will be remembered that in estuaries the tide is often asymmetrical, with a short period of rise between low and high tide but a longer period of ebb. This kind of tidal curve can be approximately reproduced or predicted by using not only the M_2, or semidiurnal tide, but also the M_4, or quarter diurnal, and the M_6, or sixth diurnal constituent. The M_4 and M_6 curves, repeated four times and six times daily, obviously do not represent any lunar or solar rhythm, as in the case of the semidiurnal tide. They merely describe the distorted wave form of the actual tide and so make possible predictions.

Based upon tidal predictions, a series of annual tide tables is published each year. In Great Britain this is the responsibility of the British Admiralty. In the United States, predictions and tide tables are prepared by the U.S. Coast and Geodetic Survey. These are indispensable to the careful navigator but also extremely useful to such diverse groups as shell collectors, harbor engineers, marine biologists, and anglers and fishermen.

The tables give the times and heights of high and low waters for each day of the year for about 180 principal ports of the world known as reference stations. The height is given above or below the same low water datum used on the nautical charts, so that the actual depth of water at any place at a given time is found by adding the height of tide to the depth marked on the chart. In the United States the reference depth is mean low water. Tables are also provided that enable tidal heights to be calculated for times between high and low water. In addition to the principal ports, or reference stations, about 5,000 other places are listed with tables showing the difference in tidal heights or times between these places and the nearest reference stations. For unlisted places a few observations with a tide pole in normal weather will serve to provide the tidal differences from the nearest reference station tides.

It has already been mentioned that in analyzing the tide, after removing all of the periodic constituent tides, there remains a non-periodic, apparently random variation, resulting from meteorological and other effects, as shown in a previous diagram (page 153). Since the tables do not include such effects in their predictions, it is obvious that under abnormal conditions the tides will not behave as predicted. Nevertheless, most of the time the accuracy is very high. At Liverpool, England, for instance, the error in the time of high tide for 94 percent of the time is less than 10 minutes and the error in the height of high tide, with a mean range of 21 feet, is less than one foot for 90 percent of the time.

Although it is not generally known, for certain purposes, tides for future years can be predicted without waiting for publication of the tide tables. This is because of the fact that at the end of every nineteen years the lunar phases repeat themselves. In effect, this means that the tidal tables for the next nineteen years will be approximately the same as those for the past nineteen years. Thus, in planning future expeditions or other marine activities in which a knowledge of tides is required, the tide tables of past years can be used. For the year 1978 the predictions from the 1959 tables would be approximately correct as to the times of high water, although there would be some error in tidal heights resulting from variations in the moon's declination and distance from earth.

It is also useful to know that when a particular place is not listed in the tide tables, its tidal data can be computed from the tables for a distant station having the same type of tide. Thus, the Hudson Strait tides are similar in nature to those of the Bay of Fundy, more than 1,000 miles away, since both have a greater variation resulting from the distance between earth and moon (apogee-perigee) than the variation between springs and neaps. Once the difference in time and the relative heights of these places are worked out from observation, the Bay of Fundy tables can be used for Hudson Strait.

Although tidal prediction is usually quite accurate, there will be times when the tides are higher or lower than the predicted tide. The observant navigator or fisherman can often make corrections for these discrepancies. In most cases they are due to changes in barometric pressure or the effects of ordinary high winds. When the pressure drops in the general

Disaster by floods as seen at the Dutch village of St. Philipsland in 1953. This was caused by a storm surge, which brought unusually high tides along with high winds, causing 20-foot waves that conspired to break the dikes. (Netherlands Information Service)

area, the sea level rises. Thus, a drop of 12 millibars, or ½ inch of mercury in the barometer, is equivalent to a rise in sea level of about 7 inches. It is less easy to forecast accurately the effect of high winds, but, in general, an offshore wind will tend to lower the sea level and an onshore wind to increase it. So, even if there were no other effects, conditions of reduced barometric pressure and an onshore wind could be expected to increase the sea level and if this happens at high tide the effect will be an abnormally high tide.

On occasions, which fortunately are comparatively rare, tidal heights are so much greater than the predicted, or astronomical, tides that they are the cause of major disasters, with great damage to property and structure and considerable loss of life. These unusual occasions are not to be confused with tsunamis, which result from earthquakes or other violent earth movements, nor with wave surges, which result from the emergence of destructive waves from distant storms. The abnormal tides we speak of are properly called storm surges. They have a murderous record. They are particularly common in the North Sea, where they are known as sea bears from the Low German *see booren*. They have occurred eight times during the last century.

The history of the Netherlands drawns attention to the so-called St. Elizabeth flood in 1421, which inundated 72 villages and killed 10,000 people. One of the earliest storm surges to be recorded with any detail also took place in the North Sea, in 1571. The extent of the damage caused by it was described by the historian Holinshed. It was a devastating storm, with heavy wind and rain. These, aided by an unusually high tide, caused widespread destruction. In Lincolnshire the seawalls, originally constructed by the Romans but neglected since the time of the Nordic invasion, could not contain the flood. Whole towns were destroyed, bridges swept away, and ships carried over the land. In 1779 high tides invaded the church in Boston, also on the east coast of England.

Although storm surges are most often reported from the North Sea, where the low-lying shores accentuate the risk of destruction of life and property, they also occur in other parts of the world. In 1876 a cyclone in the Gulf of Bengal caused a disastrous flood and the fantastic total of 100,000 victims. In the Gulf of Mexico, the city of Galveston was virtually destroyed by a hurricane flood in 1900, and in Cuba 25,000 people were killed in 1943. Galveston is particularly unfortunate, since it also sustained heavy damage from a storm surge on September 8, 1850. On September 6, the barometer dropped rapidly. On the following day, heavy swell rolled in from the Gulf. The next day, half of the streets were underwater and 6,000 people were killed. Heavy flooding also occurred during Hurricane Carla in 1961.

Most of the heavy storm surges in the North Sea in the present century have been attended with heavy death and destruction. In 1916 the dikes of the Zuider Zee were breached and vast areas were flooded. As a result of this calamity, the decision was made that led to plans for enclosing the Zuider Zee. The greatest storm surge of recent times occurred in 1953, causing great loss of life and damage both in England and Holland. At some places the water level was more than 6 feet above the predicted tidal height and remained so for 15 hours. In parts of Holland the sea level rose 11 feet above the predicted level. Heavy winds, up to 100 miles an hour in Scotland, and 20-foot waves all conspired with the surge to breach the dikes in 400 places. This surge was followed by two others, in December 1954, which was remarkable for the fact that one surge followed another, a twin storm surge. Another great storm, in February 1962, flooded the city of Hamburg.

Thanks to our modern understanding of tides and weather and to the network of tidal and weather stations, which accumulate almost continuous information, it is possible to explain reasonably well the causes of these abnormal tides. It turns out that they follow many of the principles involved in normal behavior, such as resonance.

When a depression, or low pressure area, develops over a sea, the sea level rises, as previously explained. If the disturbance is sudden, then a wave spreads out from the source, traveling at a speed that depends upon the depth of the water. Thus, in the open ocean, the wave could travel, for instance, at around 400 knots, corresponding to a depth of about 2,000 fathoms. But the speed of movement of a depression is rarely as much as 75 knots, so that the wave rapidly leaves the influence of its source and loses amplitude. In shallow seas, however, the natural speed of the wave is considerably reduced. If it moves in time with the traveling depression, it will be continually reinforced and will grow in amplitude. Similarly, changes in wind direction, if in time with the natural period of the basin, will cause resonance, and high storm surges will develop. It is also interesting to note that the storm wave not only follows the principle of resonance but also tends to follow the rotary path of the tide wave. High winds will also drive the water, so that the sea level slopes from one place to another. At one place the sea level will be abnormally low, at another abnormally high.

In the case of the Galveston flood of 1850, the wind-induced oscillations are believed to have coincided with the natural resonant period of the offshore waters, resulting in a wave that also coincided with the tidal oscillation.

The 1953 disaster in the North Sea was a more complicated phenomenon. It was primarily the result of the passage of a major depression traveling across the North Sea from the north of Scotland in an easterly to southeasterly direction. The pressure in the center of the depression was about 27.5 inches, but in the ridge behind it the pressure rose to 30 inches. This alone would generate a wave and would cause a considerable oscillation of sea level. But there was also a front associated with the passage of the depression, in which the heavy winds changed suddenly from southwest in front to northwest behind. This would also tend to develop a surge. Added to these factors, the wind blew water into the North Sea from the north, between Norway and Scotland, to an estimated extent of 200,000 million tons. This alone would increase the average level of the North Sea by 2 feet. Still another contributing factor to the 1953 disaster could be the slope of the sea surface induced by the winds. Blowing from the direction of Scotland toward the German bight at 50 knots, it is calculated that a continued wind would tilt the sea surface by one part in 300,000. Over a distance of between 400 and 500 miles this would account for a 9-foot rise in the east. The slope of the surface varies inversely as the depth of the water and the square of the wind speed.

The Dutch twin storm surges of December 22 and 23, 1954 are interesting because they fit remarkably well into the theory of resonance. The first crest arrived near the Hook of Holland in the evening of December 21. A second surge arrived on the morning of the 23rd, just about 36 hours later, but with even greater amplitude. It seems clear that a resonance effect was involved, since the second pulse was larger than the first, suggesting a reinforcement. Also, the period between the two surges, about 36 hours, coincided with the resonant period of the North Sea for this kind of disturbance.

The disastrous storm surge of 1953 led the British Admiralty to establish a storm surge warning system to predict the occurrence of abnormal flooding. The Liverpool Tidal Institute is making progress in developing systems for prediction based upon the pattern of atmospheric pressure distribution and movement related to the natural wave periods of the North Sea. The forecasting of tides has become very efficient, and great progress is being made in the prediction of violent storm surges.

13. *Ebb and Flow*

The rise and fall of the tide exhibits enormous diversity from place to place. At some places there is virtually no tide, at others there are massive movements, at some places twice daily tides, at others one tide a day. But the tidal currents, as well as the rise and fall, are equally diverse and puzzling. Why, for instance does a strong tidal flow sometimes take place at high water and a strong ebb at low water, whereas, elsewhere, the strongest current occurs midway between high and low water? Around the coasts of Britain, the flood and ebb streams generally run for about 3 hours after high or low tide. At still other places the tide may run only in one direction. There are areas with extremely weak tidal currents, and others where currents reach the awesome speed of 16 knots.

The strongest currents in the seas and estuaries are those caused by tides. Along coastlines open to the sea the tidal currents rarely exceed 1 or 2 knots, but in shallower waters they may be considerably greater. There are also a few rivers in which the tide may move upstream as a formidable and destructive wall of water, known as a tidal bore.

Part of the diversity of behavior depends on the type of tidal wave. We have already shown that standing waves, such as occur in basins, have little or no horizontal currents at the antinodes, where the rise and fall of tide is greatest. The maximum current takes place at the node, where tidal range is least, and at mid-tide, rather than at high or low water. This is characteristic of the English Channel between the Isle of Wight and Dover. In broader basins and seas, where a rotary tide develops, the water tends to move in circular or elliptical paths, except along the coast, where it moves back and forth in a linear fashion, with the strongest currents near high and low tide. Such a tide occurs in the North Sea.

In channels where tides take the form of a progressive

wave, the greatest flow of water occurs at high tide and at low tide, with slack water in between. This is because the currents move in the direction of a progressive wave at its crest and in the opposite direction in the trough.

A curious pattern of tidal currents exists off the east coast of England, where the ebb tide may continue to flow after the flood tide has begun. This apparent impossibility is due to the scouring effect of the tidal currents. The Goodwin Sands, off the coast of Kent, is pierced by a number of channels, where the tidal currents have cut a pathway through the soft bottom sediment. The ebb tide and flood tide, which tend to follow different pathways, have cut separate channels. As a result, the flood tide may begin to traverse the flood channel while the ebb tide is still running out in its separate channel. Here, as in other places, the hydraulic gradient is also involved.

Also adding to the complication of tidal currents in estuaries is the river discharge itself. Near the mouth the tide may advance against the river stream. Higher up the river, though, where the tidal currents are less, the river stream may be greater than the tidal stream. It is thus possible to have an apparent continuous ebb stream, which diminishes during the period of flood and increases during the ebb. Even when there is both ebb and flow, the ebb current will be stronger and last longer than the flood. Where the ebb current is reinforced by strong river discharge, the currents may become very powerful indeed. In the Tuamotu Islands, in the South Pacific, currents up to 10 knots develop in the Hao channel as a result of water fed into the lagoons by breaking waves and discharged through narrow channels.

The effects of river discharge and hydraulic gradients are perhaps most dramatically shown in the reversing falls at Saint John, New Brunswick, where the river enters the Bay of Fundy through a narrowing gap. At certain stages of the tide there is a turbulent fall or rapids, as the water flows down toward the Bay of Fundy. It is impossible for boats to move against the dangerous stream. As the tide rises, the incoming stream brings the water on both sides of the gap to the same level, and for a while there is relatively calm and motionless water. As the incoming tide continues to rise, however, the downstream level increases and the falls now run in the opposite direction, upriver.

Unusually strong tidal currents are sometimes developed when two basins are joined by a narrow channel. If high tide occurs at different times at the two ends of the channel, or if the tidal range is different at each end, then there will be a hydraulic gradient and the stream will flow downhill. A well-known example is the Seymore Narrows, between Vancouver Island and the mainland. At certain stages of the tide there is a 13-foot difference in level between the two ends of the channel and a current of 10 knots develops, making navigation very difficult. A similar situation exists at Deception Pass in Puget Sound. Narrows of this kind also exist in the Alaskan shipping lanes, where good-size vessels must anchor and wait for slack water.

In the Dark Ages men took very seriously the legend of the Maelstrom, a gigantic whirlpool off the Lofoten Islands of Norway. The word itself means "whirling stream" and the ancients supposed that this graveyard of ships and men led deep down into a subterranean abyss leading back to the light of day in the Gulf of Bothnia. The true facts are not quite so dramatic, but there nevertheless do exist as many as fifty powerful hydraulic currents, with associated whirlpools, off the northern coast of Norway. The most remarkable of these is an object of great awe to persons visiting Bodö. Here a narrow passage connects two fjords, and at certain stages of the tide the Saltström current runs, reaching the colossal speed of 16 knots. It carries more than 100 million tons of water during the course of 6 hours. The noise of the flow, with its attendant whirlpools, is deafening.

Another example of the turbulent flow caused by hydraulic gradients is embodied in the legend of Scylla and Charybdis. There are two rocks guarding the Strait of Messina, between Italy and Sicily. According to legend, Scylla dwelt in a cave on one and Charybdis lived under a fig tree on the other. Because Glaucus, a fisherman, loved Scylla, Charybdis in a fit of jealousy changed her into a hideous monster. Scylla threw herself into the sea and became a rock, dangerous to mariners. Thereupon, Jupiter changed Charybdis into a whirlpool. The rock and the whirlpool, twin hazards to navigation, are the origin of the phrase "avoiding Scylla only to fall into Charybdis," meaning "out of the frying pan into the fire." The rocks and whirlpool actually exist, but the tidal range in the strait is less than a foot and the turbulent currents are only 5 knots. This is modest compared to the Channel Islands, where the tidal range is large and the currents and whirlpools are notorious. Between the island of Alderney and the Cotentin Peninsula, 10-knot currents make navigation impossible at times.

In estuaries the freshwater river discharge mixes in varying degrees with the tidal flow of salt water. Where the river discharge is large compared to the tide, there will usually be a downstream flow of fresh surface water with the heavier salt water forming a wedge at the bottom and tapering off in the upstream direction. An example of this is the Mississippi River, in which the salt wedge extends as much as 100 miles when the river is low but only a mile or two when it is high. On the other hand, in rivers where the tidal flow is great compared to river discharge, the tidal motion causes mixing of the fresh and salt water and destroys the layering effect.

In some estuaries, such as Southampton Water, during rainy periods the fresh surface layer may extend to a depth of 6 feet. On the early flood, before the tidal current is flowing strongly, the surface water will be flowing at a different speed to the deeper layers. Ships drawing over 30 feet, for instance, will be dominated by the flow of the deeper waters, and appear to drift against the surface movement. A similar effect is known in the Bosphorus, where fishermen lower drogues into the deeper layers and allow the deep flowing streams to carry them against adverse surface currents.

Sometimes the tides within a bay move in one direction only, instead of reversing. This may be due to the shape of the headlands. A tidal current running parallel to the shore may be partially diverted around the bay. When the tide changes,

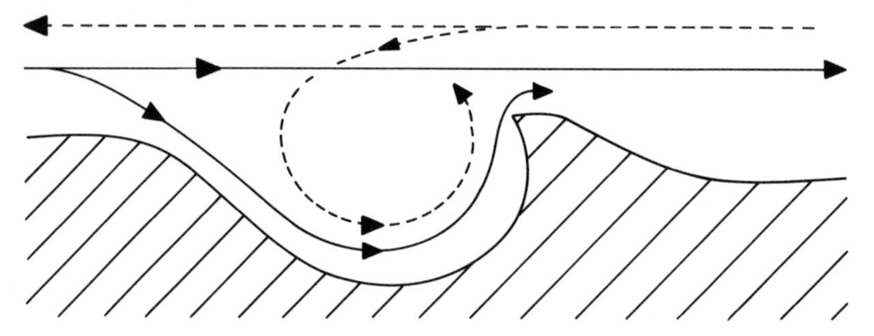

Tidal currents that never reverse occur when a tidal current running parallel to the shore is partially diverted around a bay. When the tide changes, an eddy is formed by the reverse current and thus continues to circle the bay in the same direction as the earlier current. (Richard Marra)

however, the reverse current develops an eddy, which traverses the bay in the same direction as the earlier direct current, as shown in the diagram.

An especially spectacular result of tidal currents is the bore. In the Fuchun River in China, with every tide, but especially at spring tides, a wall of water advances up the river. At spring tide it reaches the awesome height of 10 to 20 feet and travels at a speed of 16 knots. The spring tide range is around 12 feet at the river entrance, but as the bore advances into the narrowing waterway the level may rise as much as 30 feet and carry with it as much as 2 million tons of water a minute behind a white wall of bubbling foam. As described by Admiral Usborne Moore in the last century, the Chinese junks avoid the bore by building alcoves along the banks. After the passage of the bore they are then able to travel rapidly upstream in the strong current, independent of the wind. Moore wrote: ". . . no less than thirty junks swept up in the after rush and passed Haining with all sails set but their bows pointing in every direction, several proceeding stern first at a rate of ten knots toward the city of Hang Chow." In 1888 the admiral, then Captain Moore of Her Majesty's survey ship *Rambler*, made a thorough survey of the river. One of its features is the extensive area of shoals in the wide mouth between Rambler Island and immediately below Haining. An especially important observation, which gives a clue to the cause of the bore, showed that between Rambler Island and Kampu just before the bore started the sea level at Rambler Island was 20 feet higher than at Haining.

Tidal bores occur in other parts of the world. In the Tocantins River, near the mouth of the Amazon, there is a 15-foot bore, known as the Pororoca. There is also a bore in the River Hooghly in India. Smaller ones occur, usually only at spring

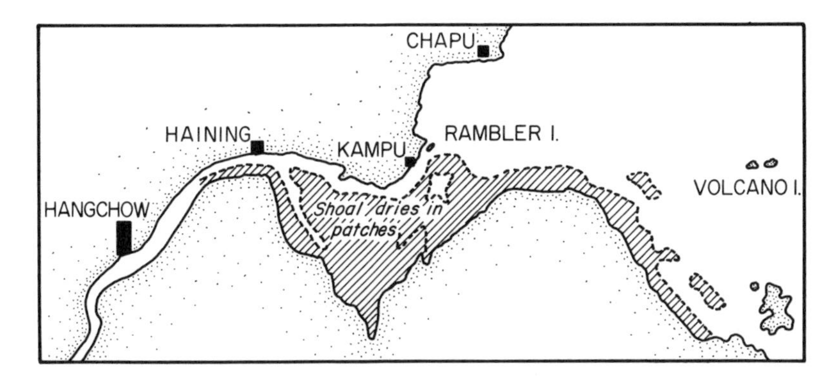

Chinese junks avoid the tidal bore that invades the Fuchun River, but on the strong current behind the bore they can ride rapidly up the river as far as Hangchow. Just before the bore starts, the sea level at Rambler Island is 20 feet higher than at the city of Haining. The shoals at the mouth contribute to the development of the bore. (Richard Marra)

A dangerous bore, the Mascaret of the River Seine, is shown in the two photographs above as it rushes through Caudebec-en-Caux. (Reproduced by permission of Dr. R. A. R. Tricker and American Elsevier Publishing Company, Inc.)

tides, in the Petitcodiac River, New Brunswick, Canada, and at Turnagain Arm, Anchorage, Alaska. In England bores are locally known as aegirs where they occur in the Trent, Kent, and Mersey. All of these are comparatively small. The largest in England is that of the River Severn, which reaches a height of 4 feet. It is noteworthy that this bore only forms when the tidal rise near the mouth is more than 30 feet.

A famous bore, known as the Mascaret, takes place in the River Seine below Rouen. This bore used to form on every tide and has been responsible for many accidents and drownings. At Caudebec-en-Caux it still reaches a height of 24 feet but is now confined to the times of the spring tides. This reduction in frequency may be due to dredging operations, which may also be responsible for its disappearance from the upper reaches.

In attempting to explain the bore, account must be taken of the features that appear to be in common. In nearly all cases there is a strong tidal gradient, the bore most often occurs at the spring tides, and there are shallow flats at the mouth. Frequently the estuary narrows as it enters the river. When the river is widened or the shallows disturbed by dredging, the bore may diminish or disappear completely. The tide rises rapidly below the shallows, the gradient increases, and when it reaches maximum the bore suddenly forms and rushes upstream into the relatively still water. It is usually true that the height of the bore is greater near the banks than in midstream. The speed of the bore is greater than that of the tidal stream, and the height of the water continues to increase for some time after passage of the bore. Behind the bore may be several lesser waves.

It has been suggested that the bore is caused by the steepening wave front of the tidal wave as it passes into shallow water. It will be remembered that this takes place in many estuaries, and that consequently the tidal rise takes place more quickly than its fall. If the speed of the crest relative to the trough in front of it continued to increase, then it is suggested that the tidal wave, instead of being miles long, would become compressed into a wall. However, this explanation is not altogether satisfactory. It seems much more likely that the bore is caused by the effect of a sudden narrowing or shallowing of the water upon the tidal current rather than upon the tidal wave. These conditions are all present in estuaries having bores.

There is a special relationship between the height of the surface of a stream in shallow water, the slope of the bottom, and the speed of the current. When the speed of the current is less than a certain amount, which depends on the water depth, then the water surface drops as it enters shallowing water. (The exact amount is $\sqrt{gd}$, where d is the depth and g the gravitational constant.) If the depth continues to decrease until the speed equals this amount, then an unstable condition arises. Any further increase in the speed of the water or the shallowing of the bed will cause the water to rise. These conditions can be seen in a mountain brook. Where the water flows over shallows, the surface will usually drop, but where it flows most rapidly over boulders, it will rise above the general surface. In the case of the bore, when it reaches shallow water, the speed is still small in relation to the depth, and the surface slopes downward in the upstream direction. But as the water shallows further or its speed increases, the speed becomes greater in relation to the depth and the water rises suddenly, traveling forward as a wave or wall of water.

No matter what the explanation of tidal currents and their diverse patterns, it is of practical importance to the fisherman,

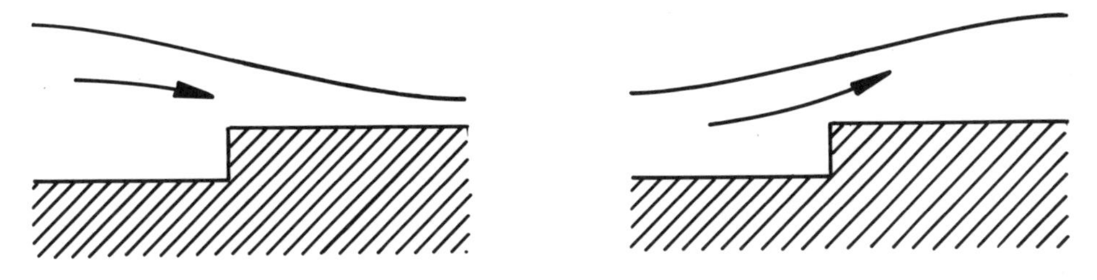

The water level will drop (left) as an incoming tide enters a shallow river or bay with a speed less than a certain critical amount. A bore is formed (right) as the speed of the water entering a shallow area increases in relation to the depth. The water surface suddenly rises and the water picks up speed, forming a rushing wave or wall of water. (Richard Marra)

navigator, or marine scientist to be able to know in advance what to expect.

The U.S. Coast and Geodetic Survey publishes current tables for a number of selected places as well as the tide tables previously mentioned. These give the predicted times and speeds of currents in a number of important harbors, estuaries, and waterways, with tables of differences that allow prediction for many other places. For the principal reference stations, there are listed for each day of the year the time of slack water when the flood begins, the time and speed of flood tide at its maximum flow, the time of slack water when the ebb begins, and the time and speed of the ebb at its maximum. For secondary stations, the tables list a time difference and a current ratio to apply to the data in the appropriate reference station. Further information is given about the average interval between the time of the moon's transit and the time of the next maximum flood. Also listed are the average speed of the flood and ebb at its maximum and the velocity of spring tides. Tidal current charts are also published for a few important harbors and approaches. These show the direction and speed of currents throughout the waterway at hourly intervals during the tidal cycle. Similar publications are produced by the British Admiralty.

14. *Man and the Tides*

The amazingly complex patterns of tidal rise and fall and of tidal currents have some practical aspects and consequences that particularly concern a large and varied group of people, including engineers, navigators, divers, shell collectors, anglers, fishermen, and scientists.

The importance of tides to navigators and harbor engineers is self-evident, as was amply demonstrated to the Germans when they attempted to plan the formidable task of invading the tidal shores of England during World War II. Those who participated in planning the successful invasion of the Normandy beaches were only too well aware of the practical effects of tides and the need to understand and predict them.

Engineers concerned with the construction of seawalls, breakwaters, quays, or embankments have long recognized the necessity of planning construction in relation to the height of extreme tides. Equally important, corrections must be made to allow for the added sea level resulting from possible storm surges. Heavy and destructive wave action will be intensified if it takes place at extreme high tides. In recent years these problems have become even more important because of the rapid increase of offshore oil drilling structures. Fortunately, records of storm tides are usually available to augment the tables of astronomical tides, so that the engineer can base his calculations upon the greatest stress the structure may be required to withstand.

Many of the older important seaports are located near the mouths of large rivers, where a knowledge of the tides is needed both for the planning of docks and for navigation, as in New York City, San Francisco, Hamburg, London, and Liverpool. One of the earliest harbors, at Cologne, Germany, was a great Hanseatic port, with a navigable channel 120 miles up the Rhine.

The ancient port of Cologne, Germany, shown in this old print, is located 120 miles inland, but vessels are able to navigate this distance on the Rhine River because of the tidal conditions. (The Bettmann Archive)

In modern times, with ships drawing in excess of 50 feet of water, docking is difficult except at high water and slack tide. In ports where the tidal range is great it is not feasible to lay alongside, so that provision must be made for tidal basins and locks to be entered and left at high tide. The special advantage of such ports as New York or Southampton is that the tide range does not exceed 15 feet. In the Thames River, where some of the freight is still landed 35 miles upstream from Tilbury docks, the ships anchor in midstream, taking advantage of the tidal currents. Barges drift with the tide from one freighter to another, loading or unloading as they go. An indirect effect of the tide in all such ports is that the tidal currents tend to scour the channels and keep them free from silting.

An aspect of tides that is assuming rapidly growing importance today is their relation to industrial and domestic waste, which is usually of such a nature that it justifies the term pollution. Even clean water discharged from the cooling systems of electrical generating plants may be harmful, because of its increased temperature, although in some cases it has been found useful in attempts to artificially cultivate shellfish and other useful fishery products. The increased temperature, if properly controlled, may increase the rate of growth of the creatures under cultivation. Nevertheless, whatever the type of pollutant entering a tidal estuary, the mixing and transporting effects of the tides must be taken into account in determining the maximum amount of effluent that should be allowed into the estuary or the probable concentration of the contaminant that may result from a plant in the planning stage.

Even in small and comparatively unimportant places the tides may assume a predominant role in the life of the community. In a number of the Scottish islands and in the Frisian Islands off Schleswig-Holstein, the regular boat services between the small communities can only navigate some of the channels at high water or at slack tide. Since this occurs 52 minutes later each day, the ferry schedule also changes daily. The result is that much of the local economic and even social life, geared to the daily coming and going of its only communication, also becomes tied to the lunar period.

The navigator nowadays has access to accurate tide and current tables. But these are not always sufficient. In many cases an understanding of the principles of tidal behavior can be extremely helpful, especially for small vessels, yachts, or fishing boats that venture into places for which accurate tidal data are not available. In places where the tide depends upon declination of the moon and there are two unequal tides each day, the navigator should be aware of the problem and be able to choose the appropriate tide. And it should be unnecessary to point out that when there is danger of going aground, the chance should not be taken on a falling tide. Nor should the risk be incurred when the tides are beginning to weaken after the springs, with the consequent delay of nearly two weeks while waiting for the next spring to provide flotation.

A curious effect of wave or tidal action may drastically alter weather conditions. In comparatively undisturbed water, a layer of warm, relatively light water tends to form over the colder, heavier water beneath like oil over water, with little tendency to mix. As explained earlier, the layer separating them is known as the thermocline. So long as this separation continues, the surface layer tends to heat up in the daytime and to cool at night, keeping more or less in step with the air temperature. But if, as a result of tidal currents, the water is thoroughly stirred and mixed, the layering effect disappears. The heat changes of the surface water are distributed to the entire water mass, and the daily changes of temperature are minimized. The result is that the air becomes warmer than the sea in the early summer and colder in the winter. The effect of warm, moist air in contact with a colder sea surface is to cause sea fogs, which commonly occur in the North Sea for this reason.

Scientists are concerned with tides from a number of viewpoints. The geologist, interested in the natural processes that scour channels and transport sediments and deposit them in certain places, has determined that some tidal currents are effective even in considerable depths. Sand waves, caused by tidal currents, have been found on the bottom in 540 feet of water off the southwest of England. In the Parrsboro Narrows of the Bay of Fundy, tidally scoured furrows have been discovered in as much as 600 feet of water. In many places the tidal currents help to carve out the coastline, building spits or shingle banks in some places, cutting passages elsewhere. In combination with waves, high tides extend the vertical range of destruction of the shore. In other places low-lying areas may be gradually reclaimed by tidal action.

Perhaps the most widespread effect of tides is their influence on marine life, particularly the forms that live between the tidemarks, although there are also tidal influences that are important to sea creatures living below the tidemarks. The intertidal zone is the most prolific area of the sea and is the home of an amazingly diverse assembly of animals and plants. Tidal behavior is therefore of great importance to those who gather or grow sedentary seafood such as oysters and mussels. It is also of special interest to the marine biologist because of the extreme variations in environment that occur between high and low tides and the various adaptations of the organisms to this.

Tides may cause large fluctuations in temperature and exposure to drying out. The alternate immersion and exposure also brings about changes in the supply of oxygen and carbon dioxide, normally dissolved in seawater, and in the supply of food materials. Exposure to heat and ultraviolet rays of the sun will also vary with the duration of exposure and depth of immersion. Below tidemarks, the effects of fluctuating water pressure must be taken into account.

Since the duration of immersion and exposure is of the utmost importance, it is not sufficient to consider only the period between high and low tide. Between the tidemarks there are all possible conditions, ranging from perpetual exposure to perpetual immersion. It is necessary, therefore, to consider the different zones between the tides.

Beginning at the highest level is the extreme high-water spring tide, the highest spring tide of the year. At this level the

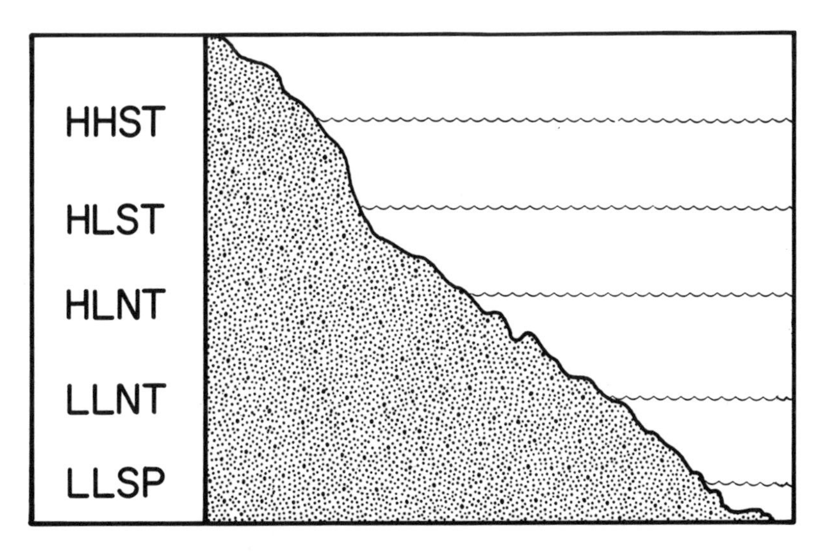

Important to marine life: Certain water levels between tides are especially significant to marine animals and plants and therefore important reference points for biologists. HHST, the high-water level of the highest spring tide. The shore above this level is never submerged. HLST, high-water level of the lowest spring tide. Below this the bottom is submerged at least once during the lunar cycle. HLNT, high water of the lowest neap tide. Below this the bottom is submerged at every tide of the year. LLNT, low-water level of the lowest neap tide. All levels above this are exposed to air during every tide. LLST, low-water level of the lowest spring tide. Below this level the bottom is never exposed. Exceptions to the above may occur during storm tides or other exceptional conditions. (Richard Marra)

inhabitants are rarely immersed in seawater, although they are subjected to spray. The next significant level is that of high water of the lowest spring tides. The rocks or shore at this level will be immersed at least once each month. High water of the lowest neap tide is the highest level at which the zone is wet at every tide. Creatures living here may include those that can only stand 5 or 6 hours' exposure to the atmosphere.

Much of the evolution of life in the sea has taken place in the shallow waters of the sea, and it is not surprising that after millions of years of exposure to the ebb and flow of the tides the behavior and life cycles of many sea animals are closely adjusted to their rhythms. Rhythmic behavior tuned to the lunar rhythms is very common, but it is not always clear whether the response is directly to the gravitational influence of the moon or to moonlight or to the tides caused by the moon. Moonlight is only about 1/500,000 as powerful as sunlight, but in some cases a continuous 24-hour exposure to light may be more important than the strength of the light. On the other hand, many of the adaptations of animals in the intertidal zone are clearly in response to the direct effects of the sun upon illumination, temperature, and humidity rather than to the moon or tidal effects.

Aristotle was one of the first scientists to notice lunar or tidal periodicity when he stated that the eggs of sea urchins increase at full moon. This belief is still adhered to by the fishmongers of Suez. Since the Mediterranean has weak tides, it is not certain whether this is a tidal response or a direct lunar response.

In more recent times considerable research has been carried out with animals possessing lunar or tidal rhythms, with some remarkable results. One of the best-known cases is that of Convoluta, a small flatworm, which appears on the sandy surface of the bottom as it becomes uncovered by the tide. A concentration of such creatures forms distinct green patches, resulting from the greenish algae that live symbiotically within their digestive tracts. As the tide comes in, the green patches disappear, the worms retiring back into the sand.

The behavior of Convoluta is understandable. The algae in its digestive tract provide food for the worm and in turn the algae probably use inorganic materials from their host. The worm only comes to the sand surface during those periods of low tide that occur in daylight. The algae need light in order to grow, so it is advantageous for the worm to expose itself to sunlight. It is equally advantageous to return to the safety of the sand when the tide turns and wave action starts. But the remarkable thing is that Convoluta continues this behavior even if placed in a laboratory tank. Even without tidal action in the tank, it will continue the same movements in time with the tide for a considerable time, until it dies for lack of sunlight.

There are many more instances of movement carried out in tune with the tides. In nearly all cases examined the movement has developed in response to the daily tidal rhythm and will continue to act in time, even when the tidal influence is removed. The innate timing system of the animal can be set in time to outside stimuli and will continue when the stimuli are removed.

Among the numerous other marine animals that show rhythmic behavior in time with the daily tides are snails, mussels, clams, and oysters. Those that crawl do so when the tide covers them. The sedentary mollusks close their shells when the tide is low to avoid the loss of moisture. When the tide returns, they resume their feeding, opening their shells and pumping water through their gills. An unusual case is the fiddler crab, which responds to both sun and tide. It turns silver-gray at night, but becomes black in the daytime as a protection against the sun. At the same time it responds to daily tidal rhythms, searching for food at low tide and resting at the flood. In the laboratory with constant illumination and no tidal movement, it will continue its behavior in time with the sun and the moon.

Other marine animals respond to the longer tidal periods, such as the spring and neap tide cycle. One of these is the grunion, a small fish that lays its eggs in the sand near the high tidemark on spring tides during the breeding season. By the time of the next spring tide the eggs are ready to hatch and are carried to sea as young larvae.

In the Pacific islands the palolo worms, which live in the coral reefs, react with great precision in the last quarter of

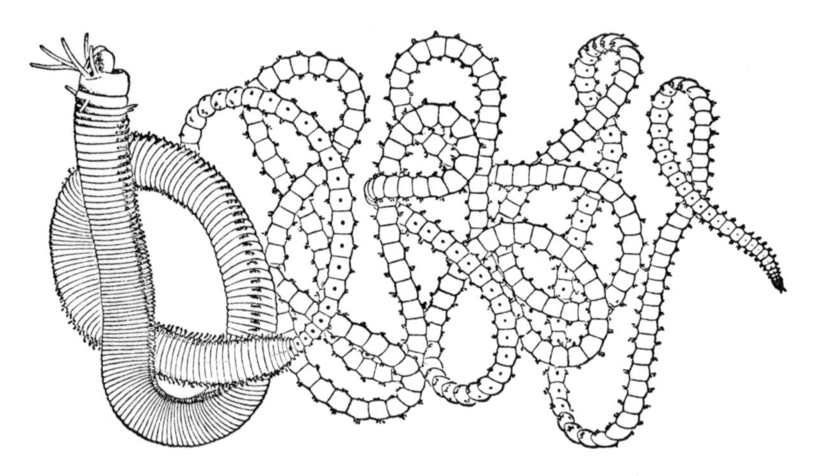

The tides and the moon influence the behavior of many marine animals. Palolo worms form a mating swarm in the marine shallows at precisely the same time each year. (W. M. Woodworth, Bulletin of the Museum of Comparative Zoology)

the moon during October and November and swarm in great numbers to the surface where they release their eggs, a highly prized delicacy in the diet of the native people as well as in that of predatory fishes. A similar worm, in the Florida Keys, chooses to do the same thing about the third quarter of the moon, between June 29 and July 28. It is believed that moonlight is at least a partial factor in this case.

An interesting example of the importance of tides to marine life, even below the tidal zone, is given by the European eel. It is well known that the adult eels leave the rivers in autumn for a long trip into the Atlantic Ocean, ending in the neighborhood of the Sargasso Sea, where they breed. After hatching in spring to early summer, the young eels travel thousands of miles to the parental rivers. The question that had scientists puzzled was, how do the eels find their way into the streams? Investigations showed that on the flood tide the eels swim toward the upper layers of water and are passively carried toward the estuaries. On the ebb tide they descend to the bottom, where they remain until the next flood tide. The result is that the tides inevitably carry them into and up the estuaries.

Anglers are well aware of the importance of tides. In some places the striped bass will only bite on the flood. The channel bass is best sought on the last of the ebb and the beginning of flood tide. In some places the weakfish can be caught only in the period immediately before high tide and the first two hours of the ebb. But in other places the same fish is caught on the last of the ebb and the beginning of the flood. Although, in

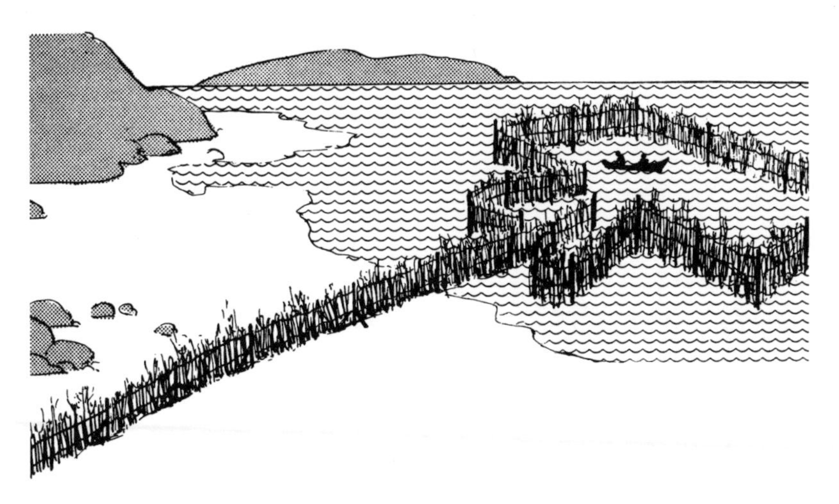

Commercial fishermen profit from their knowledge of tidal movements. When the tide flows in a certain direction, fishes moving close to the shore are diverted into this herring weir by a fixed wall. The fish pass from the wide entrance through a narrow neck. Once through the neck they are unable to leave. (U.S. Fish and Wildlife Service)

general, the flood tide seems to be preferred for most fishes at most places, there is no hard and fast rule. The only thing that can be stated with certainty is that many fishes will only bite at well defined periods in the tidal cycle.

Commercial fishermen are equally interested in tides. Even offshore trawling, in relatively deep water, can be dependent upon a knowledge of the tides. And, in shallow water, use is made of the tidal movement to trap fishes in various types of fixed nets, which are so arranged that the fish moving along the shore are diverted by a wall of netting into a trap. For the oyster farmer it is necessary to know, for instance, at what time and at what level of the tide the young oysters or spat will attach most readily to the collectors on which they are to be grown.

Thus, to many people and in many ways, knowledge of tides is not merely a matter of curiosity. It may be of the utmost importance to their activities whether these are business or recreation or the safeguarding of life and property.

15. Rivers in the Sea

To the nearshore sailor the most obvious movements of the ocean are waves and tides. The other major movements of the waters, the horizontal courses of currents, are only noticed when caused by tides or, as in offshore rips, when tides cause a dangerous increase in the steepness of waves. But the offshore navigator is well aware of major rivers in the open ocean, rivers of vastly greater magnitude than any rivers flowing from the land. Not only are there great flows at the surface, there are others below the surface, even at great depths, and still others that move the waters in a vertical direction, from below to the surface and vice versa.

The ancients could take little account of the rivers of the open ocean since their navigational methods were not sufficiently accurate to estimate currents by comparing their dead reckoning with a course made good, especially in easterly or westerly directions, in the absence of a means of determining longitude. Their early writings on the movements of the sea are mostly confined to nearshore currents, usually tidal in origin. One of the earliest mentions of offshore currents was made by the Arabic writer El-Mas' údí about the middle of the tenth century. He cited reports from sailors that currents in the Indian Ocean flow from the southwest in summer and from the northeast in winter, changing direction with the monsoon winds.

Christopher Columbus was the first to observe currents flowing westward in the tropical Atlantic during his first voyage. On September 19, 1492, while becalmed, he sounded with the deep-sea lead line. He recorded no bottom at 200 fathoms, which is scarcely surprising since the depth near his most probable position is now known to be more than 2,000 fathoms. But as the weighted end of the line sank down in the calm water, it began to pull away from the vertical. Columbus drew the correct conclusion that the lead was in motionless water below the surface and that his vessel was drifting away in a current. During the next century many more ocean currents were discovered, and theories as to the causes of currents, remarkably few at first, were gradually advanced.

Following the non-productive period of the Middle Ages, curiosity and observation were again stimulated by the Renaissance. William Bourne published a book in 1578 that dealt mainly with the tides. Like many others of his time he confused tides with ocean currents, so that we find him believing the westerly movement of the ocean observed in the Atlantic tropics to be a tidal current produced by the moon's pull on the water.

Thevet's *La Cosmographie Universelle* in 1575 suggested that the currents in the Straits of Florida were the outflow of the Mississippi and other rivers. Later it was found that this could account for a mere one-thousandth part of the Gulf Stream system. Both able scientists of their day, Johannes Kepler and Vasenius in the following century advanced the theory that westward flowing waters near the equator result from the eastward rotation of the earth, which leaves the waters behind, just as a passenger in a rapidly accelerating vehicle finds himself hurled backward. Isaac Vossius wrote an entire volume, *De Motu Marium et Ventorium*, in which he explained that the tropical heat attracts the ocean toward it in a mountain of water just as the sun draws up the morning mist into vapor. As the tropical sun moves west, this great ocean bulge follows it, until it is turned northeast by the north coast of South America. Georges Fournier in 1667 neatly reversed this theory by suggesting that the sun evaporates water beneath it, forming a huge hollow into which water flows irresistibly as the hollow follows the sun's path.

One of the first scientists to realize the importance of winds as the driving power of currents was Athanasius Kircher toward the second half of the seventeenth century. This was affirmed a century later by Benjamin Franklin, who was convinced of the importance of the trade winds blowing water westward across the equator. François Arago, in the middle of the nineteenth century, stressed the effects of the unequal heating of the sea in high and low latitudes, whereby warm currents might be caused to flow from the equator to the poles. He also stressed a most important factor, the effects of a rotating earth, a factor that had already been used to explain the deflection of the trade winds toward the west. This factor, the Coriolis force, was named after the scholar who originally drew attention to the deflecting effects of planetary motion. But it was not generally applied to an understanding of ocean currents until 1903, when the pioneer oceanographers J. W. Sandstrom, Björn Helland-Hansen, and V. W. Ekman invoked it.

Early in the twentieth century it became generally accepted that two principal driving forces keep the currents in motion and a number of factors direct and modify their courses. The driving forces are both derived from the sun's energy, either directly through solar energy given to the sea or through release of the same energy from the ocean to the atmosphere and the subsequent action of wind upon water. These directive and modifying forces interact in such a fashion that only rarely may currents be described in terms of one factor alone.

The direct effect of the sun is to cause an unequal heating of the surface waters. This in turn causes changes in the density, inasmuch as water becomes lighter when heated in the tropics and heavier when cooled at the poles. The density also changes because of variations in salinity. As a result of solar heat and winds, a loss of water by evaporation in excess of that added by rainfall may increase the salinity and therefore the density, while an excess of rainfall over evaporation decreases it. The formation of ice also increases salinity. The general result is that polar waters become dense, and the waters near the tropics become lighter. The heavier polar waters therefore tend to sink below the surface as the lighter tropical surface waters spread out toward the poles. This obviously implies that there is a compensatory return flow of deep water and an uplift of water at the equator to equal the sinking. These movements are known as a thermohaline circulation, since the density distribution depends upon both heat and salinity.

The major driving force of the surface circulation of the

ocean, however, is not thermohaline circulation but rather wind stress. It is therefore necessary to know something about atmospheric circulation, which in turn also depends upon the unequal heating of the earth's surface by solar radiation.

The earth's surface, as a whole, receives 47.5 percent of all the incident short-wave radiant solar energy, or sunlight. This is returned outward as long-wave radiation. But the incoming radiation is not equally distributed. Where the sun's rays fall perpendicularly upon the earth's surface, the maximum radiation per unit surface is received. But with increasingly high latitudes, the sun's rays become more and more oblique and so are spread more thinly over a wider area.

Thus, at the time of the equinox, radiation is received vertically at the equator with maximum effect, but at the poles the incident angle is almost horizontal and the energy received is practically zero. At other times of the year, with the changing declination of the sun, the zones of maximum and minimum radiation simply move north or south of the equator by 23½ degrees.

If there were no heat loss, the earth would become extremely hot. Since there is no significant increase in the temperature of the earth, there must be a loss of heat equivalent to the gain. This loss, or back radiation, is in the form of long-wave, or heat, energy. But whereas the incoming radiation varies from a maximum at the equator to zero at the poles, the back radiation varies relatively little. If there were no redistribution of heat upon the earth's surface, there would be a great excess of incoming radiation at the equator and a net loss of radiation at the poles. As the polar regions cooled down, their back radiation would decrease until equilibrium was reached, and at the equator the temperature would increase until its back radiation equaled incoming radiation. The polar regions would become extremely cold and the equator extremely hot. The mechanism that prevents this situation from occurring is a redistribution of heat energy by means of atmospheric transfer, the wind system, and oceanic transfer, the currents. The wind system is the major factor in bringing about the circulation of the ocean.

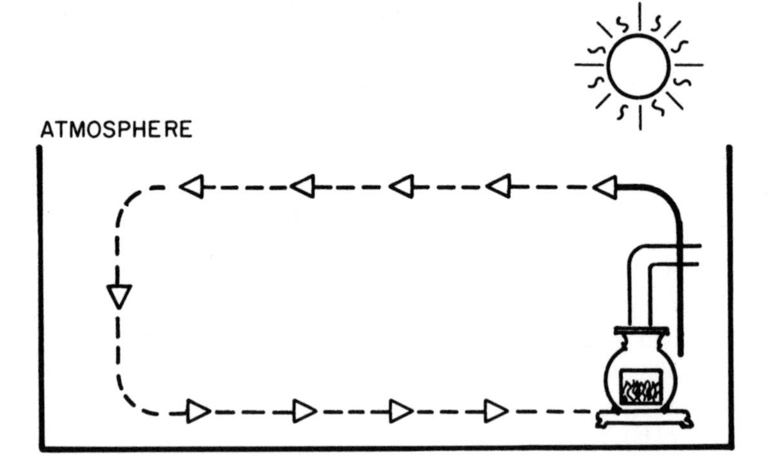

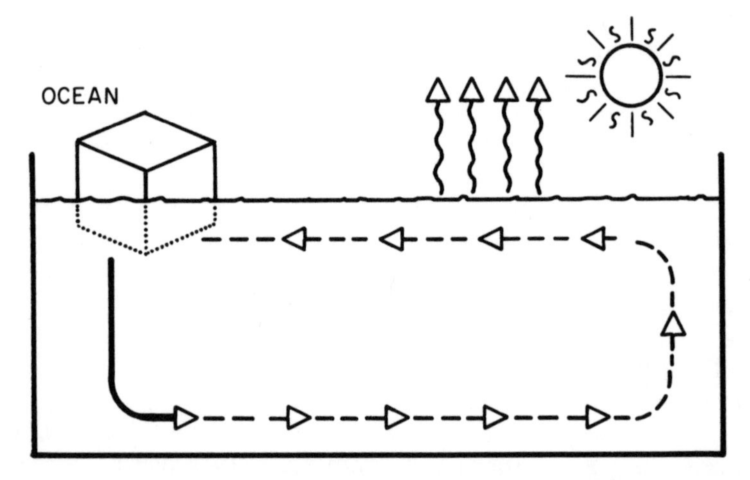

The sun is the only constant source of energy on planet earth. It drives the powerful wind circulation of the atmosphere and, both directly and indirectly, the movements of the oceans that have prevented them from being stagnant pools, unable to originate or sustain life. In the upper figure, the sun's heat, partly absorbed and transferred by the ocean water to the atmosphere in the equatorial and tropical regions, causes air to rise. Cold air flows at surface level from the poles to replace it and a circulation is set up. Air travels at higher altitudes from equator to poles and returns nearer the surface. In the lower figure, in the ocean, the polar regions suffer a net loss of heat, so that cold, saline, and therefore heavy, water sinks. Warmer, less dense, water from the tropical and equatorial regions flows in to replace it and a vertical circulation develops, with deeper water flowing from poles to equator. This very much simplified model is the thermal circulation of the atmosphere and the thermohaline circulation of the ocean.

In the tropics a considerable amount of heat is transferred from land and sea to the atmosphere. Part of this is by conduction to the air immediately in contact with the surface, and part by convection, as the air warmed at the surface rises and is replaced by the cooler air from higher latitudes that flows near the surface to replace it. Heat is also transferred to the atmosphere by evaporation at the ocean's surface, since large quantities of heat are required to change water to vapor.

On a non-rotating earth, the heating of the earth's surface and the subsequent transfer of heat to the atmosphere would bring about a general uprising of warm air in the tropics and an inflow of cold surface air from the poles to replace it. The circulation would be completed in the upper levels of the atmosphere by a reverse compensatory flow from the tropics to the poles. This simple system of a single, closed cell in each hemisphere is greatly modified as a result of two major factors, one of which is the Coriolis force, resulting from the rotation of the earth, and the other, the unequal heat properties of land and sea. The Coriolis force has already been used to explain amphidromic, or rotary, tides. It is of even more general importance in the study of ocean currents and of paramount importance in the study of atmospheric circulation.

To an observer fixed in space, an object set in motion and allowed to move freely actually continues in a straight line, but what is a straight line in space appears as a curve to an observer upon a rotating set of coordinates (the earth's lines of latitude and longitude). The movement of an orbiting earth satellite demonstrates this. If the satellite is in polar orbit, it

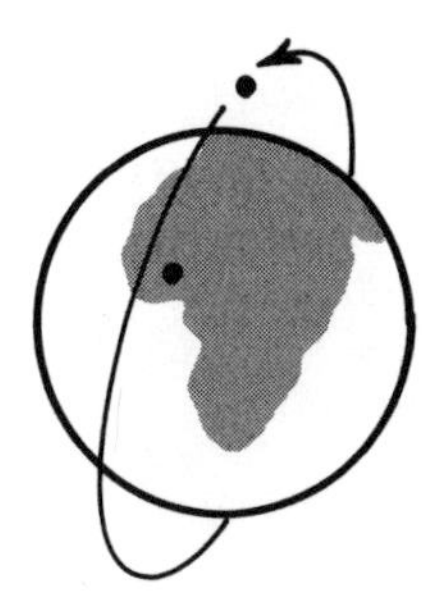

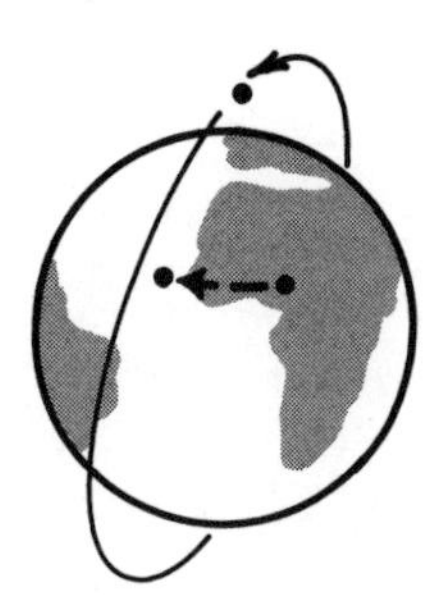

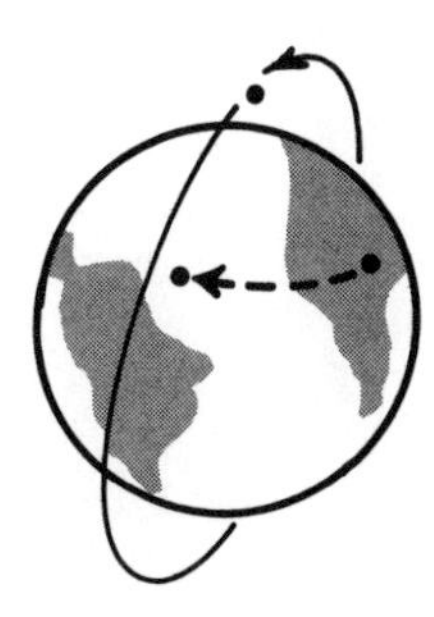

An example of the Coriolis force. The satellite orbits the earth in a path that remains fixed in space. But the earth is rotating to the east upon its axis. The satellite therefore appears, *to an observer at a fixed spot on earth, to be deflected to the right in a spiral that curves to the west in the Northern Hemisphere.*

continues to move in the same stationary orbital plane, but the earth revolves underneath it. As a result, each time the satellite crosses the equator it will appear farther to the west, depending upon the rate of rotation of the earth. To an observer on earth, however, it appears to spiral to the right. A simpler demonstration may be easily conducted at home by taping a plain cardboard sheet to the turntable of a record player, which is set in rotation. If a straight edge is held stationary above the turntable in any direction and a pencil drawn along it in a straight line at a steady rate, the pencil will record a curved track upon the sheet. Since the turntable is rotating in a clockwise direction, corresponding to the planet's Southern Hemisphere as seen from the pole, the

The Coast Guard cutter Pontchartrain *battles rough weather on Ocean Station Baker. Ocean station ships stay at fixed locations, where they gather weather and ocean information, provide navigational aids for ships and planes, and perform search and rescue duties, if needed. (U.S. Coast Guard)*

track curves to the left. On the earth's surface this effect is slight but nonetheless existent. Its magnitude is in proportion to the speed of the moving object and increases with increasing latitude, being zero at the equator and maximum at the poles.

The Coriolis force is allowed for in the aiming of naval guns. A shell fired in middle latitudes with a velocity of 1,800 miles per hour, for instance, would be deflected more than 200 feet from its path by the time it reached a target 20 miles away. Without a Coriolis correction such a missile, with the middle of a destroyer as its target, would completely miss the enemy.

Although no actual deflecting force is involved, it is convenient for practical purposes to consider that the movement of an object across the earth's surface is deflected from a straight line on the earth's coordinates by a fictitious force. This imaginary force, the Coriolis force, is proportional to the rate of the angular rotation of the earth. It is also proportional to the sine of the latitude, that is, it is zero at the equator and increases with increasing latitude until it becomes maximum at the poles. It is also proportional to the velocity of the current, increasing as the current speeds up.* All of these factors may be demonstrated upon the record player.

Returning to a simple atmospheric circulation, in the ab-

* At any point at rest on the earth's surface there are two principal forces acting. The gravitational force of the earth is directed toward the earth's center. The force resulting from the earth's rotation is a centrifugal force directed outward, in much the same fashion that a weight twirled on the end of a cord imparts an outward pull to the cord, or the way an object on a rotating turntable is flung off. This centrifugal force is perpendicular to the axis of rotation and is proportional to the square of the earth's speed of rotation and inversely proportional to the distance from the axis. The two forces combine to produce a force directed downward. The earth is flattened toward the poles, as a result of centrifugal force, and the combined force is perpendicular to the earth's surface at that point. If an object at this place is set in motion in an easterly direction, its speed of rotation becomes greater than that of the earth and the centrifugal force is increased. This additional force thus developed is perpendicular to the axis of rotation. It may be resolved into two components. One acts vertically and has no horizontal deflecting effect. The other acts tangentially along the surface of the earth and is directed southward. This is the deflecting Coriolis force or acceleration in the Northern Hemisphere. If the object moves to the west, the centrifugal force of the stationary condition is reduced by a force acting toward the earth's axis. The tangential component of this is directed to the north. Thus,

sence of the Coriolis force, on a uniformly heated, nonrotating earth there would only be heavy, irregular convections of rising air scattered over the earth's surface, each accompanied by cumulus clouds and thunderstorms. Between the cumulus areas, a compensating flow of cold air would descend to the surface. But as a result of unequal heating at the equator and cooling at the poles, a single cell would be set up in each hemisphere with rising air at the equator and sinking air at the poles, a movement toward the pole in the upper atmosphere and movement toward the equator at the surface.

The Coriolis force deflects the surface air moving away from the north pole so that it moves to the right, that is in a westerly direction. As the resultant east winds pass into warmer areas of lower latitudes, the air becomes lighter and rises. At the higher altitudes it returns toward the pole along its original path. Thus the single pole-equator cell of a nonrotating earth is short-circuited.

In a similar way, the rising air at the equator short-circuits the single cell. As it rises and moves toward the poles, it is deflected to the right, or in an easterly direction, becomes cooler, sinks, and forms another cell. Between the equatorial and polar cell, the air adjusts to the movements in a third cell. If one could picture the three cells in three dimension, the air would be seen to move as through adjacent left- and right-

---

in each case the motion of the object is subject to an acceleration to the right of its motion.

If the stationary object is set in motion in a northerly direction, its angular momentum, or "spin," remains unchanged, as when a wheel on well-lubricated bearings is set in motion, it would, except for friction, retain its momentum and continue to revolve at the same speed. But the moving object by traveling north in the Northern Hemisphere decreases its distance from the earth's axis. Since its angular momentum remains the same, the speed of rotation increases to compensate for the reduced radius of rotation, just as a stone twirled on a string increases its rate of revolution when the string is shortened. This causes it to move easterly at a faster rate than the surface beneath it. In other words, it receives an acceleration to the right. Conversely, an object moving south increases its distance from the axis of rotation and consequently rotates more slowly. In this way it receives a deflecting acceleration to the west, or right. Thus, an acceleration to the right of the path of a moving object exists whether moving in a latitudinal or longitudinal direction. In any other direction, since it can be resolved into latitudinal or longitudinal components, the same holds true and the magnitude of the Coriolis force is $2\omega \cdot \sin \Phi \cdot V$, where ω is the angular rotation of the earth, Φ is the latitude, and V the speed of the moving object.

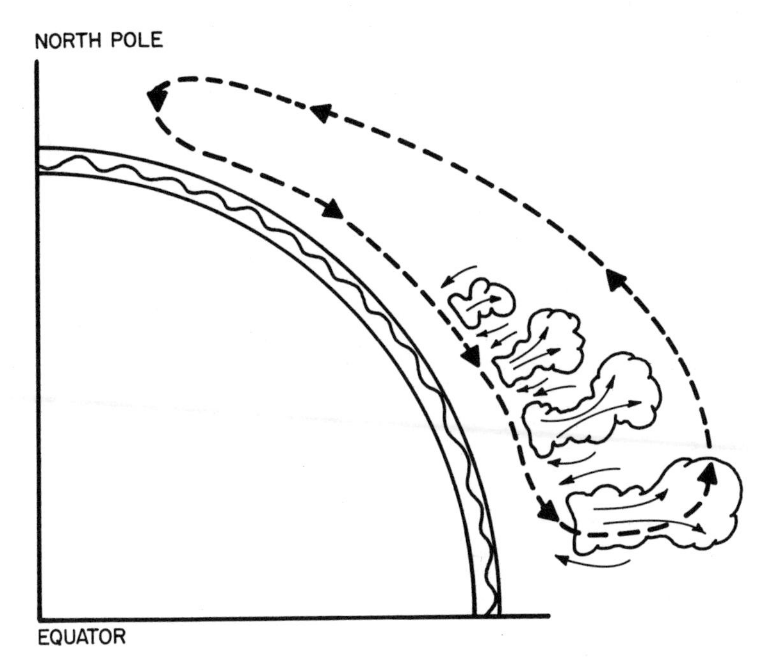

Since the sun's heat is applied mainly in low latitudes, on a nonrotating earth there would be heat convection and rising air in the vicinity of the equator, northerly winds near the ground, southerly winds aloft, and descending air in the cool polar regions.

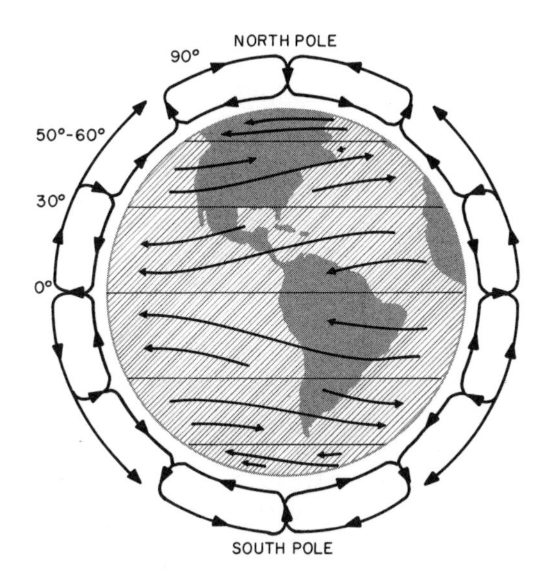

Because of the earth's rotation, the simple single convection cell of the preceding diagram is broken into three cells. The surface northerly winds moving south from the North Pole are displaced to the right as the polar easterlies and so do not complete the single pole-equator cell of a nonrotating earth. The high altitude northward moving winds from the equator are also diverted to the right, or east. Accordingly, two cells develop, with a third cell between them.

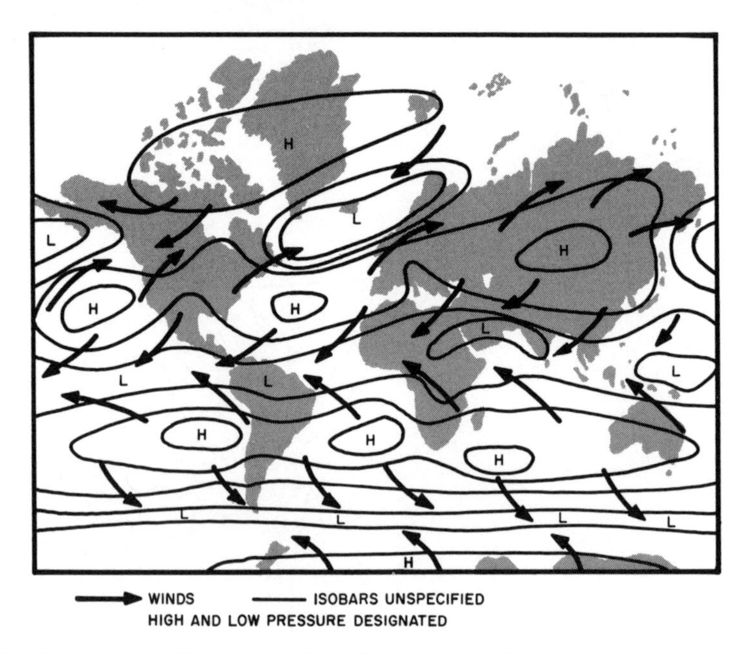

Heating and cooling on land and sea are different. As a result, the simple concept of alternate high and low pressure bands circling the earth is modified in actuality and the bands are broken up into high and low pressure areas. The diagram shows the distribution of these areas and the winds associated with them.

hand threaded screws moving parallel to the latitudes. In the Northern Hemisphere the surface air moves in an oblique direction, alternately from northeast, southwest, and northeast. It ascends at the equator and at about 60 degrees latitude and descends at the poles and at 30 degrees latitude. The wind belts are the polar easterlies, the westerlies of the mid-latitudes, and the northeast trades of the tropics. The intermediate belts of calm air are the doldrums at the equator and the horse latitudes at about 30 degrees north, where in the days of sailing ships there always existed the danger of being becalmed.

Thus far the effects of unequal heating and of the Coriolis force account for three main bands of wind in each hemisphere, separated by alternate areas of high and low pressure. The isobars, or lines, connecting areas of equal pressure, would lie parallel to each other and to the lines of latitude. The areas of descending cold air are high pressure, those with ascending warm air are low pressure. These belts of winds and pressure areas move north and south with the sun.

Another factor that influences the wind system is the distribution of land and sea. Water has a high heat capacity, that is, it requires a considerable amount of heat to increase its temperature. Land, on the other hand, has a low heat capacity, so that the same amount of heat will bring about a greater increase in temperature of the land than of an equivalent amount of sea. Conversely, the same amount of heat loss will result in a much greater drop of temperature on land than in the sea. Hence, the land becomes hotter than the sea in summer and cooler in winter. Since warmer areas tend to become areas of ascending air and therefore of low pressure and colder areas tend to become areas of descending air and of high pressure, the land will develop high pressure areas in winter and low pressure areas in summer. The consequence of this is that instead of a simple system of continuous parallel belts of high and low pressure there is actually a series of alternating high and low pressure cells, which change with the seasons. The isobars now form circular systems, similar to the bottom contour lines on a nautical chart.

The winds that develop around the major high and low pressure areas at sea do not flow directly from the high pressure areas to the low pressure areas. Because of the Coriolis force, as they move from high to low pressure they turn to the right in the Northern Hemisphere and to the left in the Southern Hemisphere. If one then looks at the weather maps of the daily newspapers, which show the isobars, or lines connecting places of equal pressure, it will be seen that the winds around low pressure areas do not move across the isobars directly into those areas but move instead in a direction almost parallel to the isobars in a counterclockwise direction around the low pressure areas in the Northern Hemisphere and in the reverse direction in the Southern Hemisphere. Around a high pressure area the winds move in a clockwise direction in the Northern Hemisphere and counterclockwise in the Southern Hemisphere.

The circulation of air around a low or high pressure cell is best explained by first considering air movement without the Coriolis force. A low pressure area has less weight of air above it than a high pressure area and is the equivalent of a depression in a solid surface. The force of gravity acting along the downward pressure gradient would pull the air directly toward the center of the depression, so that the wind would cross the isobars at right angles. If the Coriolis force is considered, in the Northern Hemisphere it will act to the right of the wind direction and deflect it until it begins to cross isobars obliquely. The wind continues to be deflected to the right of its new direction until it finally moves in a counterclockwise direction parallel to the isobars, and equilibrium is reached. The Coriolis force and the gravitational force along the pressure gradient are now balanced, being equal and opposite.

In the case of a high pressure cell the reverse situation occurs. The air flowing outward from a high pressure area is deflected to the right and so moves in a clockwise direction. This type of flow is called geostrophic, meaning "earth-turned," when the only forces involved are the Coriolis force and the pressure gradient and the flow is steady. When either wind or water flow is braked by friction forces, the Coriolis force is lessened and the wind direction becomes more oblique with respect to the isobars. What has been said of the wind is equally true of ocean currents in their relation to the density distributions and sloping surfaces of the sea.

While the foregoing account of the winds provides an understanding of the major wind systems, it must be remembered that it describes only average conditions, derived from studies extending over a period of time. There are, however, fluctuations of various kinds and local winds of varying intensity and direction as the high and low pressure cells grow or diminish and move their positions, not only with the sun's declination but also at lesser irregular intervals. The presence of land will bring about winds directed offshore in the evening, as the land cools more rapidly than the neighboring sea. In the morning, as the land warms up, the local winds will tend to blow onshore.

The unequal heating of land and sea not only causes daily and local wind fluctuations but also in some cases very pronounced seasonal ones. An especially well-defined example is that of the winds in the northern part of the Indian Ocean. During the Asian summer, the land becomes much hotter relatively than the adjacent sea surface. As a result, the atmosphere over the land becomes a low pressure area, and wet southwest monsoon winds blow in over the land. In the

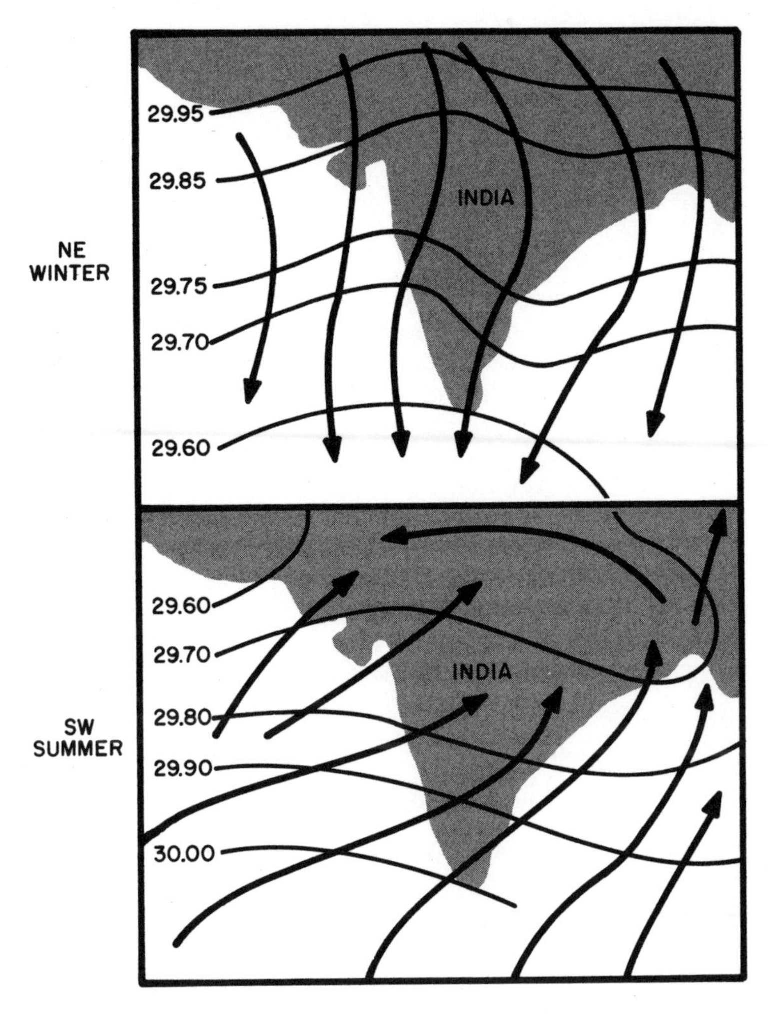

Seasonal changes are largely due to the movements of the sun and the meteorological equator, northward in our northern summer and southward in the winter. The resultant change in wind distribution is well illustrated by the seasonal reversal over the Indian Ocean. The Asiatic winter monsoon is a dry wind (top). The summer monsoon is wet (bottom).

winter, the land becomes relatively cold and develops a high pressure area. The winds now reverse, and the dry northeast monsoon winds blow from land to sea.

Another seasonal change that has profound effects is due to the declination of the sun, as it moves from a position vertically over latitude 23½ degrees north in the Northern Hemisphere summer to latitude 23½ degrees south in the Northern Hemisphere winter. The result of this is that the whole system of winds tends to move north and south, since they are in large part generated by unequal solar heating. The division between the Northern Hemisphere winds and the Southern Hemisphere winds may be called the thermal equator, which shifts north and south of the geographical equator with the movement of the sun.

16. *The Current Engine*

The effect of thermohaline circulation in driving surface currents is considerably less than that of the wind systems. Since thermohaline circulation is also involved in vertical and subsurface movements, it will be considered later. In addition to the driving power of the winds, it is also necessary to consider the effects of inertia, whereby a major current system will continue to flow, driven by its momentum, in spite of fluctuations in the wind system. Also to be considered is the local piling up of water masses by wind action, which results in a downhill flow from higher to lower sea levels. Another factor is the fluctuation of sea level resulting from local differences in barometric pressure. In general it may also be said that wind-driven currents alter the density distribution of the seas and that density differences, in turn, modify the currents. There is, accordingly, an interaction between the wind-driven and the thermohaline circulation.

When the wind blows across the surface of the ocean its pressure fluctuations exert vertical stresses upon the surface that bring about wave motion. But at the same time, horizontal stresses exert a drag that causes a wholesale drift of the waters. This is of minor importance in the case of temporary winds, but in the case of the more or less permanent wind belts, such as the polar easterlies, the mid-latitude westerlies, and the northeasterly trades, considerable momentum is built up and the wind drifts generated in this manner become permanent currents. These currents persist, with some variation in location and velocity, in spite of fluctuations in the winds. They are carried along, as it were, by their momentum.

Wind-driven currents, or drifts, do not usually travel in the direction of the wind itself. Drift currents in the Northern Hemisphere, in deep water and in the absence of land barriers, are deflected by the Coriolis force in a direction to the

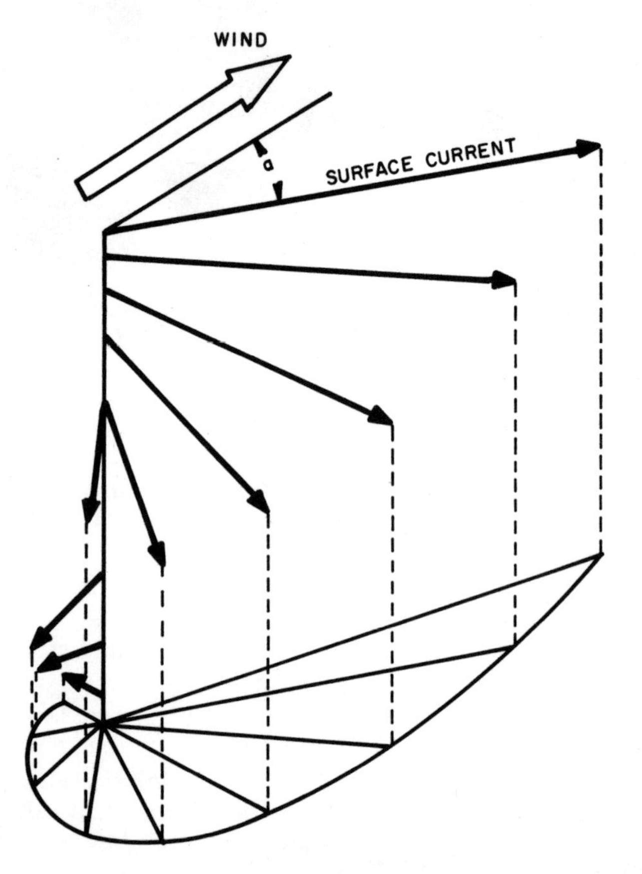

Wind driven currents at the surface flow in a direction that forms an angle to the right of the wind direction (in the Northern Hemisphere), because of the Coriolis force. The surface current imparts some of its motion to the underlying water, which in turn is deflected to the right. At successively greater depths there is further deflection and loss of speed until at the greatest depth of the current there is a very weak current flowing in a direction opposite to that of the wind. The total water flow is in a direction at right angles to the wind.

right of the wind. The current direction at the surface is at a maximum angle of 45 degrees but usually at a lesser angle to the direction of the wind. The surface layer imparts a portion of its energy to the layer of water immediately beneath, which in turn is deflected to the right of the surface flow but at a decreased velocity. At successive depths the flow is deflected still more to the right and further diminished in velocity until at a depth of about 50 fathoms it is moving very slowly in a direction opposite to that of the wind. The total transport, that is, the movement of the water column added together, is at right angles to the wind. These changes in direction and velocity with depth are known as the Ekman spiral.

When a wind drift is generated in shallow water and there is no adjacent land, the Ekman spiral develops less strongly. The surface current flows at an angle of less than 30 degrees to the wind and the total transport at less than 90 degrees. In very shallow water, considerably less than 50 fathoms, the current and transport are in the direction of the wind itself or at a very small angle.

Sailors and yachtsmen should remember that major currents, such as the Gulf Stream, continue to flow because of their momentum, but the effects of local wind changes will superimpose a current, as defined in the previous paragraphs, in addition to or subtraction from the average predicted current.

If there were no land obstruction, the currents resulting from a uniform wind belt would run a course encircling the earth. However, when winds in parallel belts blow in opposite directions or at different speeds or when land boundaries exist, as they do in most parts of the world, the currents are directed inward toward a central area, piling the warm surface water up above the general level of the ocean. This high pressure water corresponds to the high pressure air cell, as described in Chapter 15, and the water behaves in much the same way as the air. For example, in the North Atlantic, currents resulting from the westerly winds tend to be deflected to the right toward the southeast, and the trade winds tend to drive currents toward the northwest. As a result, the warm surface water forms a pool in mid-ocean, the Sargasso Sea, with the surface at a higher level than the surrounding colder ocean. The surface water runs downhill and outward but is deflected to the right until the gravity force resulting from the sea slope is in equilibrium with the opposing Coriolis force. The result is a current that encircles the warm center in a clockwise direction on a sloping surface that separates the warm pool from the surrounding cooler ocean.

The average surface circulation of the world's oceans, neglecting seasonal and other fluctuations, follows closely upon the wind systems. In each ocean, except where interrupted by

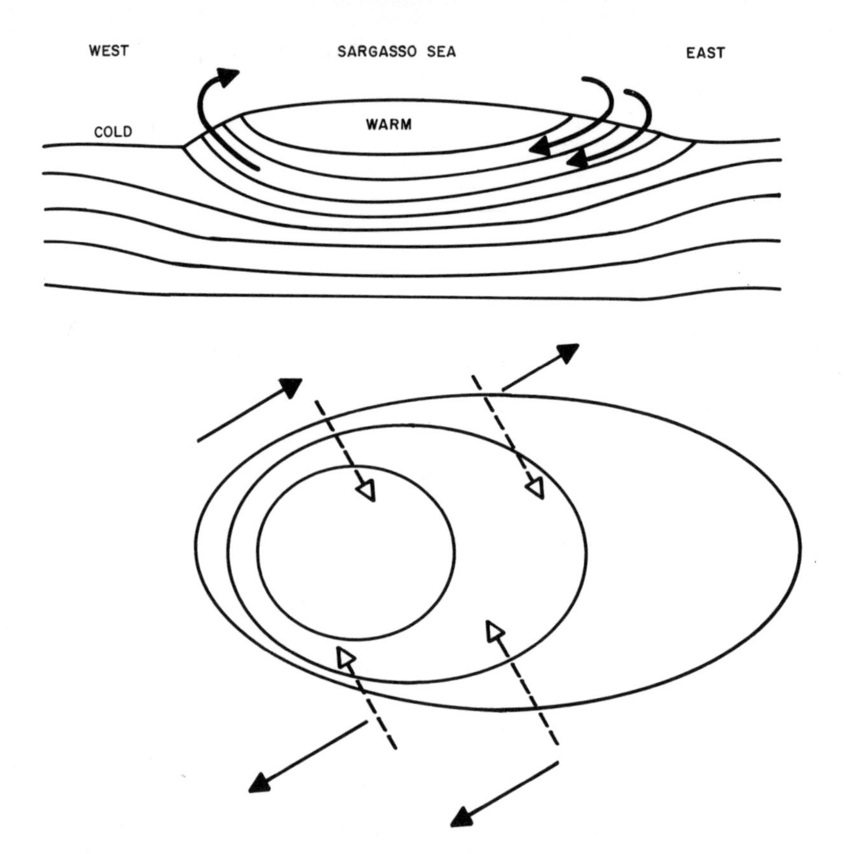

The result of the zonal wind distribution over the Atlantic, for instance, is to pile warm water up at the center of the ocean in the Sargasso Sea. Just as in the case of the circulation of air around a high pressure area in the atmosphere, the water flows in a clockwise manner around the center of the ocean with the warmer water to the right and the colder water to the left, and a relatively steep temperature gradient across the stream.

land masses, there are a clockwise circulation, or gyre, and a counterclockwise gyre in the Northern Hemisphere. In the Southern Hemisphere these gyres run in the opposite circular direction. Near the equator the diminishing Coriolis force makes it possible for some of the westward flowing northern and southern equatorial currents to flow without forming a gyre. Although part of these waters, as they reach the western boundary, join the subtropical gyre, the remainder forms a compensating countercurrent, which moves eastward between the westward flowing northern and southern equatorial currents. The remaining major current is the West Wind Drift, a continuous westerly flow of water around the Antarctic continent that acts as part of the southern Atlantic subtropical gyre.

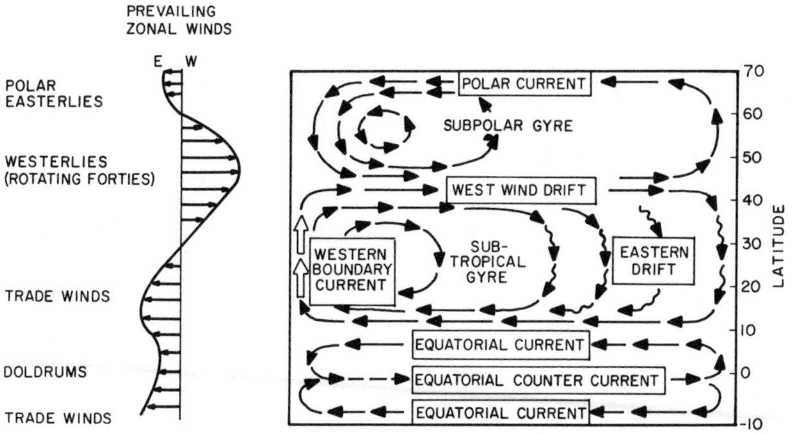

The wind systems of the major oceans and the resulting water circulation. On the left of the diagram the prevailing winds are shown to alternate from east to west, beginning with the polar easterlies and followed first by the westerlies and then by the trade winds. The resultant water circulation consists of a subpolar counterclockwise gyre and a subtropical clockwise gyre. At the equator the north and south equatorial currents moving to the west are separated by an equatorial countercurrent. The western arms of the gyre are narrow and fast flowing, the eastern portions broad and slow. (After W. Munk)

This generalized description is subject to modification of various kinds in different oceans. The West Wind Drift is driven only partly by the westerly winds. The water of greatest density is found near the Antarctic, so that the water slopes upward in the direction of the warmer surface water to the north. The water tends to flow from the warm to the cold water but is deflected to the left, causing an eastward drift. The drift is slow and, in fact, it has been shown that at the rate of about 8 miles a day it takes about four years to complete the circumpolar path. Nevertheless, the total volume of flow is about 200 million tons per second. Submarine ridges deflect the flow to the north in various places, and in the South Atlantic, a portion bends north as the Falkland Current. Some of this water may join the subtropical gyre. In the absence of land, the free passage surrounding the Antarctic continent allows the West Wind Drift and the Circumpolar Current to develop in place of a subpolar gyre.

In the South Atlantic Ocean the principal currents of the subtropical gyre consist of the Benguela Current, which flows northward along the west coast of Africa; the South Equatorial Current, which runs obliquely across the equator; and the Brazil Current, which runs southward along the coast of Brazil. The counterclockwise gyre is completed at the point where the Brazil Current joins the West Wind Drift.

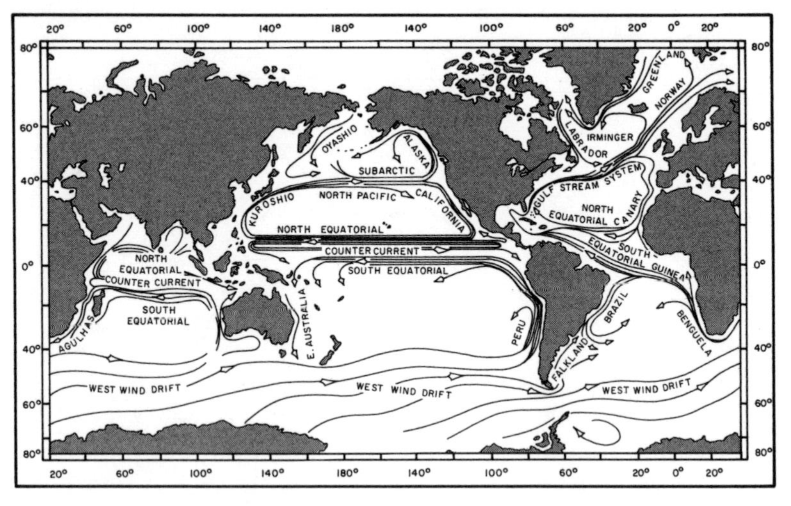

The major surface currents of the world oceans, based upon average values. (After W. Munk)

The Benguela Current is a slow drift of cold water, with the coldest water to the right-hand, or coastal, side. There the water is further cooled by an upwelling of colder water from beneath the surface. It transports about 16,000,000 tons per second. As it leaves the coast of Africa, it becomes the South Equatorial Current, which moves eastward at a rate of about 20 nautical miles a day. Because of the uneven distribution of land masses, the thermal equator is displaced to the north of the geographical equator, so that the South Equatorial Current actually crosses the equator into the Northern Hemisphere. The effect of the Coriolis force north of the equator is to deflect the northern part of this current, transporting 6,000,000 tons per second, into the North Atlantic, while the remaining southern portion is deflected southward against the bulge of Brazil to become the Brazil Current. Of the original 16,000,000 tons per second transported by the Benguela Current and the South Equatorial Current, only 10,000,000 tons per second enter into the Brazil Current. The net loss of 6,000,000 tons of surface water per second to the North Atlantic must obviously be replenished by means of a sinking of water in the North Atlantic and a return across the equator by subsurface southerly movements. This will be described later when dealing with subsurface currents.

In the North Atlantic Ocean, the subtropical gyre is well developed. This is a continuous flow with the Sargasso Sea at the center. Its principal constituents are the Gulf Stream System, the Canaries Current, and the North Equatorial Current. The latter is a broad, shallow, westerly drift of waters, which is a continuation in the tropics of the southwesterly currents off the coast of northwest Africa. The gyre moves in a wide band between 10 degrees and 30 degrees north with its maximum velocity in the southerly portion, where the average speed is about 15 nautical miles daily.

As the gyre approaches the West Indies, part of it enters the Caribbean Sea, while the remainder passes north of the West Indies as the Antilles Current. The trade winds, driving the equatorial currents westward through the Caribbean, pile up the waters in the Gulf of Mexico. This increased sea level is an important factor in the Gulf Stream System. The waters leaving the Yucatán Channel, impelled by a difference in water level between the Gulf of Mexico and the Florida Straits of about 8 inches, amount to an average of 33,000,000 tons per second. This part of the Gulf Stream System has a maximum surface speed in excess of 3½ knots somewhat to the west of the middle of the straits.

North of the Bahama Islands the two great streams the Florida Current and the Antilles Current reunite to form the Gulf Stream proper, which begins to turn eastward toward European shores. At first it remains fairly compact but then meanders from side to side, often forming loops, which may become separated from the stream itself as rotating pools or eddies surrounding a cold-water core. Also, the stream may divide into a series of separate filaments, which continue on an independent path for awhile, only to disappear as they lose their momentum and as new filaments appear.

The total transport of the Gulf Stream includes the 33,000,000 tons per second of the Florida Current and about 16,000,000 tons per second of the Antilles Current, together with much slower and diffuse currents to the east of the main flow, for a maximum that has been estimated at more than 80,000,000 tons per second.

As the water flows eastward it loses its compact nature and becomes the North Atlantic Drift. Part of it completes the circle by moving as the broad, diffuse Canaries Current along the coasts of France, Spain, the Azores, and North Africa.

Part of the North Atlantic Drift moves north between Green-

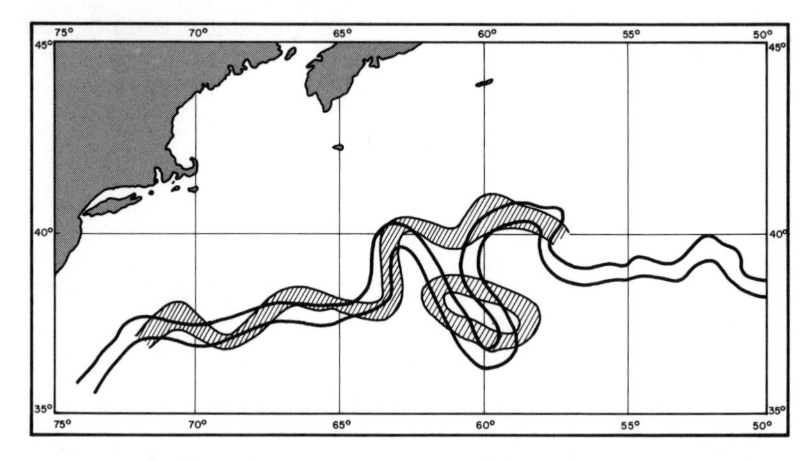

Sailors, for instance yachtsmen racing from Newport, Rhode Island, to Bermuda, may often be confused by surface water temperatures into believing that they are in the Gulf Stream when they have actually entered an entirely separate eddy. In the diagram, the original position of the Stream is shown as an unshaded pathway, with a large loop extending to the south. The shaded pathway shows the situation some time later, when the loop has become entirely separated. It is now an eddy of warm water moving in a counterclockwise direction and enclosing a cold core of water that was originally north of the Stream.

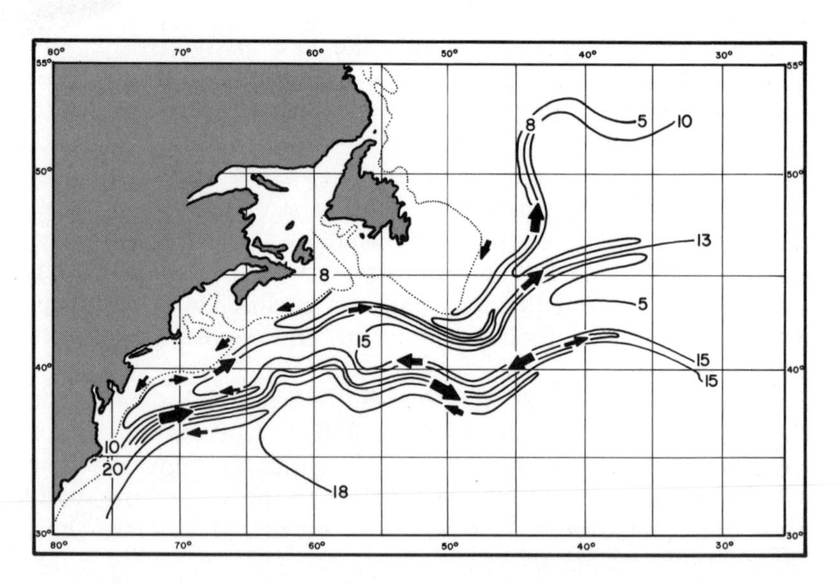

The picture of the Gulf Stream as a single narrow current with clearly defined boundaries is an oversimplification. The above diagram, showing surface isotherms, or lines of equal temperature, indicates that independent current filaments branch out from time to time, only to die away. The water flow follows the isotherms in a general way. (After W. Munk)

land and Scotland across the Wyville-Thompson Ridge into the Norwegian Sea. This is the current that moderates climate and allows Norwegian ports in the same latitudes as the Greenland ice cap to remain open. It also makes Iceland habitable. Another part of the North Atlantic Drift curves back around the south of Greenland as the Irminger Current. The water entering the Arctic Ocean as the Norwegian Current returns south along the coast of Greenland as the East Greenland Current and is joined by the Irminger Current as it moves northwest along the west coast of Greenland. The water flow is balanced by the cold Labrador Current, which returns Arctic water to the lower latitudes, part of it as a cold tongue of water that runs between the Gulf Stream and the coast of North America until it disappears below the surface. The Norway Current and the Greenland Current and also the

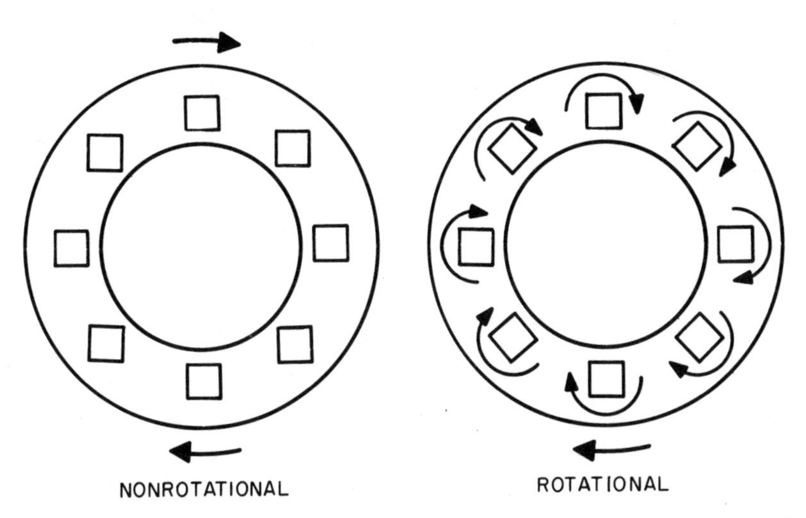

Nonrotational and rotational flow. The diagram shows a stream of water flowing in a circular closed gyre, as seen from above. The squares represent individual columns of water within the stream. In the left-hand figure the columns move around the gyre but do not themselves rotate, just as a pencil does not rotate when held vertically in the hand and moved along a horizontal circle on a sheet of paper. In the right-hand figure the columns not only move around the gyre but also rotate on their own axis—they have acquired a spin. This is rotational movement.

WIND ALONE

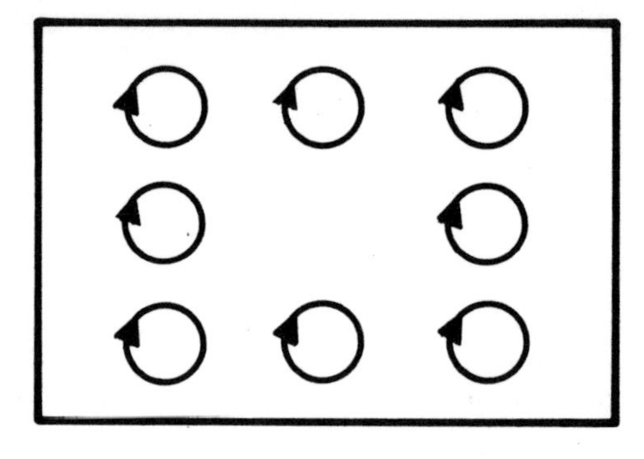

SYMMETRICAL

WIND PLUS
CORIOLIS FORCE

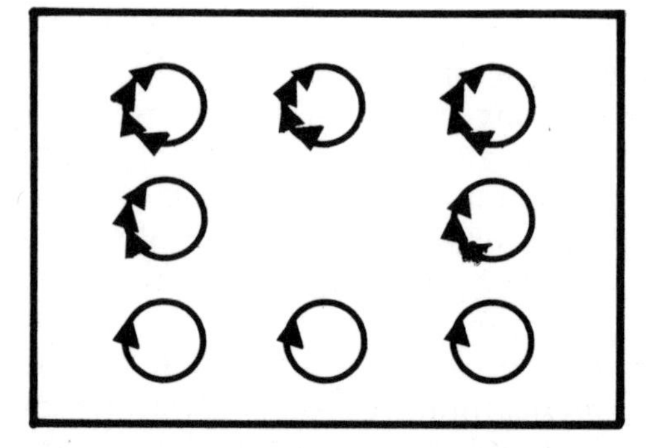

SYMMETRICAL

WIND PLUS
CORIOLIS FORCE
MINUS BRAKING

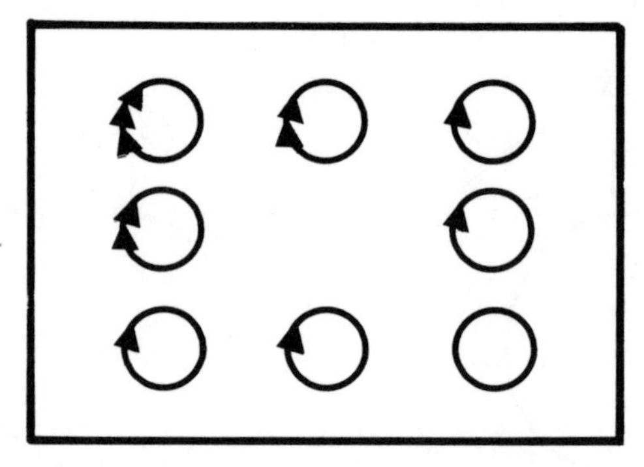

ASYMMETRICAL

The above diagrams show the development of vorticity, or spin, in an ocean circulation gyre. In the upper figure, the zonal winds around the ocean (see diagram page 201), mainly westerly in the north and easterly nearer the equator, not only drive the water around the gyre, but impart a spin to individual water columns (see preceding diagram). The winds alone produce essentially the same amount of spin in all parts of the gyre.

In the second figure the Coriolis force is added. This increases on the western margin as the water moves into higher latitudes. There is no further change as it moves to the east, but the spin is now decreased as the water travels toward the equator and the Coriolis force decreases.

In the third figure the braking effects mentioned in the text reduce the spin during travel along the western and northern parts of the gyre. The net result is a stronger spin along the western margin than along the eastern margin of the ocean.

Irminger and Labrador Currents may be considered as irregular subpolar gyres.

The western part of the North Atlantic gyre is the narrow, fast-flowing Gulf Stream System, consisting of the Florida Current and the Gulf Stream proper. The eastern part is the broad, slow-moving Canaries Current. This pronounced asymmetry is due to an asymmetry in the vorticity, or spin of the water. If it is imagined that the ocean is divided into a large number of individual water columns standing side by side, then each column would have a rotary movement of its vertical axis. In rotational flow each individual column of water rotates.

The westerly winds and the northeasterly trades together set the gyre in motion and give the water a clockwise spin that is independent of latitude. In addition to this, the earth's rotation adds clockwise spin to the water as it moves north in the Gulf Stream System and the Coriolis force increases. In the absence of any other forces the spin would remain constant as the water flowed to the east in the North Atlantic Drift. In the south-flowing Canaries Current, however, the water loses clockwise spin as the Coriolis force decreases. So far, asymmetry has not developed. However, the system is kept in balance and reaches a steady state as a result of the development of braking forces. These include fluid friction on the outer borders of the current and also the energy losses resulting from meanders, eddies, and the separating current filaments previously mentioned as occurring in the streams as it begins to flow east. The result of these is to reduce clockwise vorticity. So, while vorticity increases in the west, but to a lesser extent than it would without braking forces, it decreases in its easterly flow and continues to decrease in the Canaries Current. The net result is an asymmetry, with greater net vorticity in the west than the east.

The pronounced asymmetry of spin is reflected in the width of the current. The spin of adjacent water columns or cells may be added together to obtain larger and larger columns and so obtain the entire picture. When all columns have similar spin, the sides that touch, moving in equal and opposite directions, cancel out. But when the westerly cells have more spin

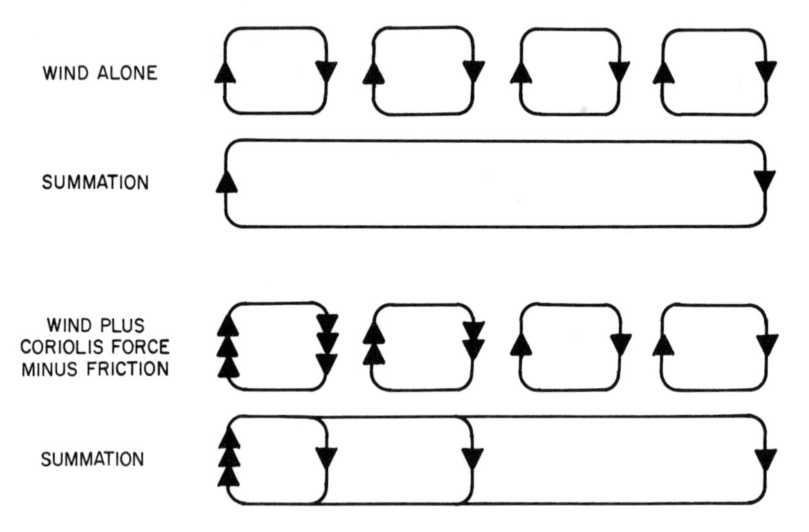

Stronger vorticity, or spin, along the western edges of the ocean gyre than along the eastern edges accounts for the narrower, stronger flow in the west. This pronounced asymmetry is explained in the diagrams. The spin in adjacent columns of water may be added together to obtain the spin over the entire gyre. In the upper figure, where the wind spin alone is involved, the spin is the same at all points from west to east. When added together, the equal and opposite movements in adjacent sides of columns cancel out, leaving a single symmetrical cell. In the lower figure, with greater spin in the west, the adjacent columns do not completely cancel out, but leave a strong concentrated spin in the west and a diffuse one in the east.

than the easterly cells, the result is a narrow, fast current on the west and a broad, slow current on the east.

The movement of the thermal equator north and south with the declination of the sun alters the distribution of wind stresses and this, in turn, affects the circulation of the current gyres. During the course of a year, therefore, the rate, as described in a previous chapter, of water transport varies. The seasonal range in transport of the Gulf Stream proper is about 15,000,000 tons per second, with the maximum occurring in March and the minimum in November. In the Straits of Florida the maximum is in July and August and minimum in November. In sailing between Miami and the Bahamas the maximum current set to be allowed for is about 3 knots in July and August and about 2¼ knots in November. The reason that the Gulf Stream maximum (in March) does not coincide with the Florida Current maximum (in June) is probably due to a shift in the relative amounts of Equatorial Current water that flow in the Florida Straits and in the Antilles Current.

A well-marked feature of the Gulf Stream System is the upward slope of sea level from Florida northward, which

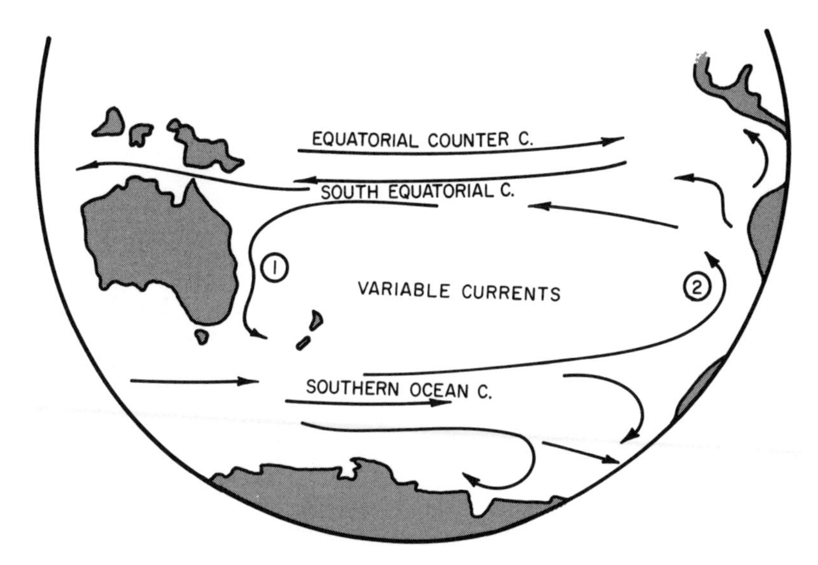

Surface currents in the southern Pacific Ocean. (After Cotter)

amounts to about 8 inches higher off New Jersey and about 12 inches off Halifax, Nova Scotia. This piling up of water may be caused by the pressure of the southwest wind. It results in a southwesterly countercurrent along the northeastern shores of the United States, with an eddy between this and the Gulf Stream. Farther south there are a number of eddies between the Gulf Stream System and the shore. These may account for the well-defined series of embayments along the coast.

In the North Pacific Ocean the two principal current gyres are similar to those of the Atlantic. The Kuroshio, moving northeast off the shores of Japan, is the equivalent of the Gulf Stream. It continues as the North Pacific Drift in an easterly direction and then toward the equator as the California Current, which is the counterpart of the Canaries Current. The circulation is completed by the westward flow, the North Equatorial Current. A partial subpolar gyre consists of the northwesterly Alaska Current and the southwesterly Oyashio. Except for the much greater expanse of the Pacific Ocean, the same general principles apply as in the Atlantic.

In the South Pacific the East Australian Current, which is in a comparable situation to the South Atlantic Brazil Current, runs south until it meets the West Wind Drift. The well-known Peru, or Humboldt, Current, carrying cold water, turns northward along the coast of Chile until it joins the South Equatorial Drift, thus completing the gyre. As in the Atlantic, there is a countercurrent between the North and South Equatorial Drifts.

As in the Atlantic, the warm tropical waters of the Pacific, moving toward the pole along the westerly margins and then to the east, influence the climate of the northern parts of the continental western shores, so that British Columbia has a more equable temperature than places at the same latitude in Eastern Asia. By the same token, the south-moving waters of the California Current are cooler than the average for the latitude as they bathe the California coast. Similarly, in the South Pacific, the Peru Current brings cooling waters to the coasts of Chile and Peru.

The Peru Current is interesting because of the occasionally disastrous effects of seasonal changes in the position of the thermal equator. During the northern winter, between Janu-

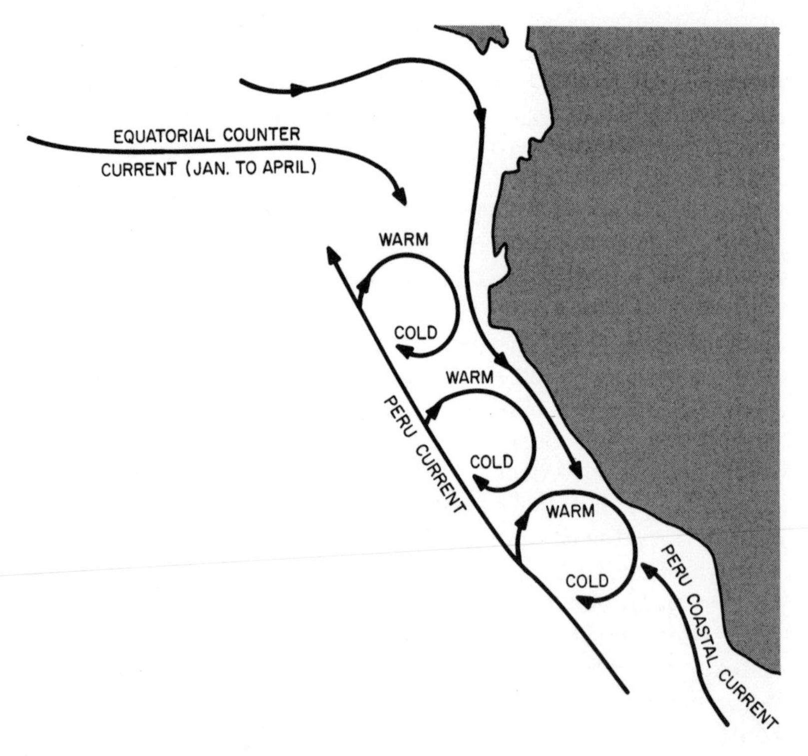

Surface currents along the Pacific coasts of Ecuador, Peru, and northern Chile. (After Cotter)

ary and April, the Equatorial Counter Current is no longer deflected mainly to the north but shifts southward, and part of its warm waters, meeting the coast, is deflected along the coast in a southerly direction inside the Peru Current. Not only is the Peru Current normally cold from its Southern Ocean origin, but the prevailing winds, blowing offshore, cause the surface water to move outward and cold water to well up from below. When the warm waters from the Equatorial Counter Current run south along the coast, the conditions become untenable for cold planktonic organisms, so that some destruction occurs, leading to the starvation of the

Surface currents in the Indian Ocean. (After Cotter)

higher organisms that feed upon the plankton. This warm inshore current is exceptionally well-developed about once every seven years. On these occasions it is known as El Niño, since it occurs at Christmas. It causes considerable destruction of fishes and of the birds that feed upon the fishes. This effect is further intensified by heavy rains in the Equador area, also resulting from the shift of the thermal equator. The rains cause an increased freshwater flow from rivers to enter the ocean and add to the unsuitability of the water for organisms used to a cold saline environment. When the destruction is particularly heavy, the hydrogen sulfide generated by decomposition may blacken the white lead paint used on ships. This result has given the name Callao Painter to the more severe El Niño phenomena.

The Indian Ocean, which is mostly in the Southern Hemisphere, understandably has a permanent counterclockwise sub-

Surface currents in the northern Indian Ocean during the S.W. monsoon.

Surface currents in the northern Indian Ocean during the N.E. monsoon.

tropical gyre south of the equator only. This consists of the West Australian Current, the South Equatorial Drift, the Mozambique Current, which flows south along the east coast of Africa, and the West Wind Drift.

Because of the monsoon winds, the northern part of the Indian Ocean experiences a complete reversal of ocean currents. During the northeast monsoon period, from November to April, both a westward flowing current and an Equatorial Counter Current are developed, the latter west to east. During the southwest monsoons the countercurrent continues, but the westward flowing current is reversed and the entire flow becomes clockwise.

While the surface currents are usually noticeable only when they affect the navigation of a ship, there are other currents, as noted earlier, that are completely invisible beneath the surface, some at the very bottom of the ocean, some at intermediate depths, but moving in broad bands and very slowly. There are many reasons for knowing this to be so, even without means of measurement. In the first place, analyses of samples of deep water show that they may retain dissolved oxygen, though to a lesser extent than surface water. Since in deeper layers the animal life removes oxygen and no plant life exists in the total darkness to replenish this, the water would ultimately become completely deoxygenated unless it were continuously, however slowly, replenished through contact with the upper waters. Stagnation does, in fact, occur in small, enclosed seas such as the Black Sea where there is no mechanism for vertical circulation.

In the North Atlantic, with a net influx of 6,000,000 tons per second of surface water from the South Atlantic, the volume and therefore the sea level would continually increase if there was not a mechanism for returning subsurface water south and for replenishing this water by the sinking of the surface water. There are, in fact, three major zones where water is withdrawn from the surface to lower layers, from which, eventually, similar amounts will reach the South Atlantic directly or indirectly to equalize the water budget.

At the entrance to the Mediterranean, as mentioned in discussing internal waves, there is a shallow sill over which a surface current enters the sea from the Atlantic Ocean, with an equivalent current beneath it returning to the Atlantic. Because there is an excess of evaporation over precipitation and river drainage in the Mediterranean, the surface waters be-

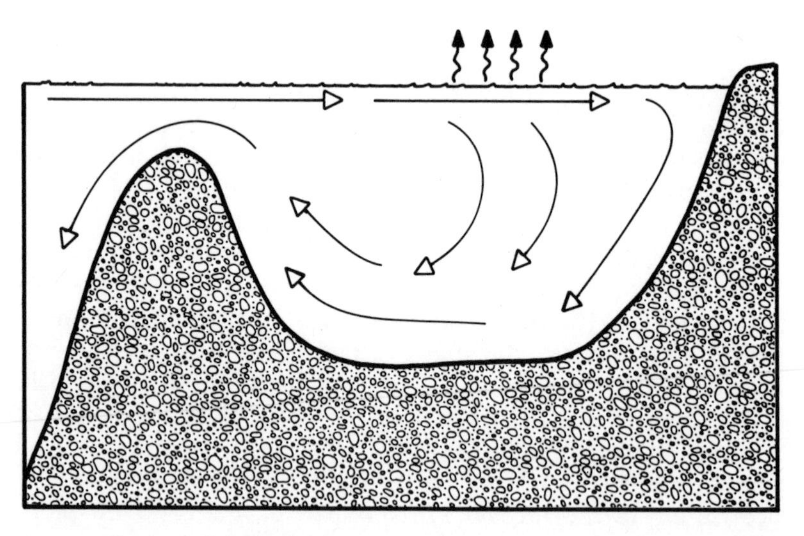

Section through the basin of a sea with sill at entrance and an excess of evaporation over precipitation and runoff. This is exemplified by the Mediterranean. In such basins, lighter water flows in from the ocean at the surface and the saltier heavy basin water flows out beneath. The water in the bottom of the basin is constantly replenished.

come saltier and heavier than the lighter Atlantic waters, which ride in above them while the deeper, heavier waters pass out and sink to an intermediate depth in the Atlantic, where they are surrounded by waters of similar density. This circulation keeps the bottom layers of the Mediterranean well flushed and oxygenated. The Black Sea has a reverse situation. Excess precipitation keeps the surface waters light, so that the surface waters flow out over the sea. The bottom waters are therefore not flushed with new oxygenated surface water, but with old subsurface water from the Mediterranean. The result is that the Black Sea bottom water accumulates nutrient material, but becomes relatively poor in oxygen, and is nearly devoid of life.

The net sinking of water from the Mediterranean is about 2,000,000 tons of water per second. Surface losses of similar magnitude take place where the Cold Labrador Current and where the Greenland Current converge with the Gulf

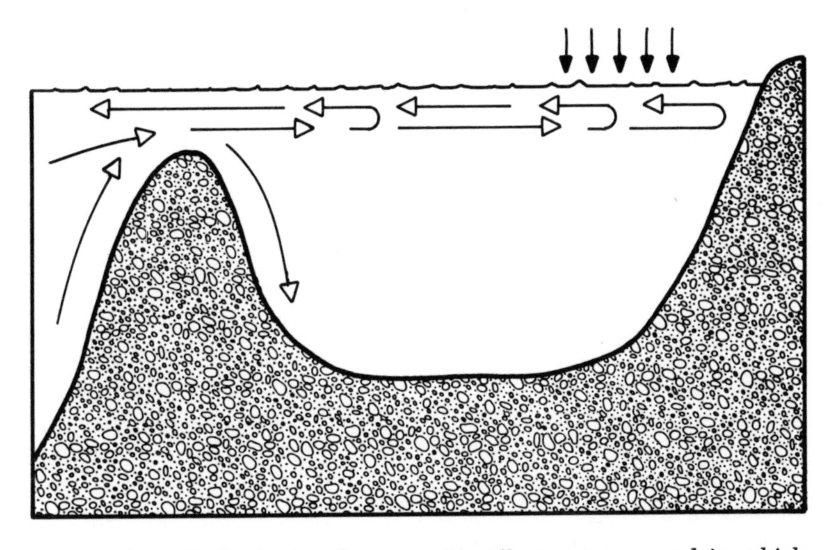

Section through the basin of a sea with sill at entrance and in which precipitation and runoff exceed evaporation. This is exemplified by the Black Sea. Lighter water at the surface runs out into the neighboring sea and is replaced by a flow beneath the surface. But since there is no outflow of deep water, the water at the bottom of the basin is stagnant.

Stream. Altogether, these account for most of the 6,000,000 tons of surplus at the surface.

An important source of cold and dense bottom water is the edge of the Antarctic Continent where ice is formed. In the formation of ice, the water freezes, leaving behind it most of its salt content in the remaining liquid, which therefore becomes more saline.

Water at the surface may also sink when it becomes heavier because of cooling, a process that occurs mainly in the polar regions. It also becomes heavier whenever there is an excess of evaporation over precipitation. This brings about thermohaline circulation, a general drift of surface waters toward the poles, a sinking at higher latitudes and corresponding rising of water in lower latitudes. Another mechanism that promotes vertical movements is caused by the convergence or divergence of surface currents. When currents move toward each other an excess of water would accumulate unless it sank below the surface. Where currents move apart, the void between them is filled by water that moves in from below.

Another cause of rising or sinking water is a wind blowing more or less parallel to the shore. The resultant drift in the Northern Hemisphere is initially to the right of the wind. Thus, if the wind is blowing to the south along the western coast of a continent, the drift will be away from the shore. There will thus be a deficit of water near the shore, which is met by an upwelling from the subsurface layers. The surface slope will be in balance with the Coriolis force, and a surface current will flow at a small angle to the right of the wind, almost but not quite parallel to the coast. If, on the other hand, the wind blows toward the north, water will pile up toward the coasts, and there will be a sinking movement. The equilibrium current will again be slightly to the right of the wind direction.

Whenever water sinks below the surface it will continue its downward movement until it reaches a depth at which its density is similar to that of the surrounding water. Its salinity and temperature may differ, even though the resultant density is the same, so that the sinking water may retain its identity for a considerable time. Eventually, the water mixes with the adjacent layers and again rises to the surface, either through a divergence or by thermohaline circulation. However, this is a very slow process and it may be as much as 1,500 years before the deep bottom waters complete their excursion from the surface to the bottom and back again. Water reaching the surface today may have sunk to the bottom long before Christopher Columbus crossed the Atlantic.

The water at the very bottom of both the Atlantic and Pacific oceans mainly originates in the Antarctic, although some Arctic water enters the western Atlantic basin.

The details of subsurface circulation are still not completely known. This is partly because the slow, deep movements are difficult to measure directly. Most of the information is derived from such indirect measurements as the distribution of temperature and oxygen values and by theoretical considerations of the effects of convergence, divergence, and the deflections resulting from the Coriolis force.

Subsurface waters may be defined by the depths to which they extend. Thus, the cold saline waters from the edge of the Antarctic ice descend to the bottom of the ocean and are known as Antarctic Bottom Water. Cold waters sinking in the northern North Atlantic are known as North Atlantic Bottom

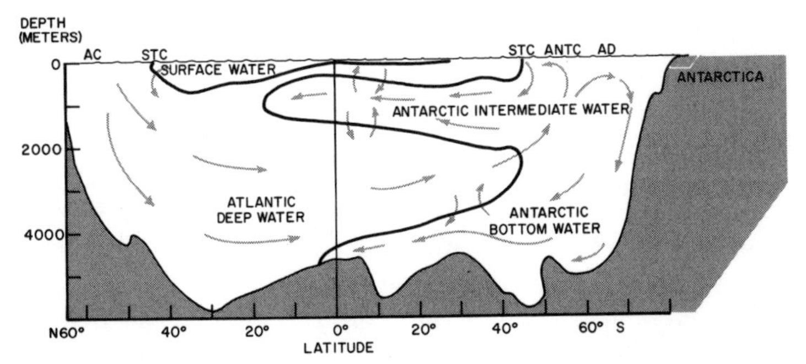

Vertical section showing subsurface circulation in the Atlantic Ocean. The principal water types are shown, with lines indicating their approximate boundaries. AC, Arctic Convergence; STC, Subtropical Convergence; ANTC, Antarctic Convergence; AD, Antarctic Divergence. Surface water sinks at the convergences. It also sinks where cooled by ice formation, as at the edge of the Antarctic Continent, or where salinity is increased because of evaporation. Water rises at divergences and in regions where offshore winds cause upwelling. (After A. Defant)

Water and also North Atlantic Deep Water. The deep waters flowing to the south of the equator extend above the layer of Antarctic Bottom Water. Sources of sinking water at intermediate depths are derived from two principal zones of convergence, the Antarctic Convergence and the South Atlantic Subtropical Convergence. The warm but saline Mediterranean Water also sinks to an intermediate depth.

The deep currents are believed to be concentrated generally in movements along the western boundaries of the Atlantic, Pacific, and Indian oceans. In the Atlantic Ocean the flow is southward from subpolar origins across the equator. In the Pacific the flow is northward across the equator. In the Indian Ocean the flow is also northward. From these major flows, the water spreads out to the east and toward the poles. Moving along the margins of the Antarctic is the Circumpolar Current. Much closer to the surface, in the Atlantic Ocean just north of the equator, there are vertical movements of water between the westward flowing equatorial currents and the easterly countercurrent, rising in the divergent area and sinking in the convergent area.

There remains to be mentioned one other type of subsurface current that is not part of the wind or thermohaline circulation. Estimated to reach the fantastic speed of 55 knots, turbidity currents, as they are called, are due to the slumping of soft sediments that have accumulated on submarine slopes. When disturbed, the unstable sediment rolls down the slope, carrying water with it as a mass of liquid mud. The best-known example of this occurred in November 1929, when an earthquake shook the floor of the Grand Bank off the coast of Newfoundland. It was estimated that the resultant turbidity currents traveled 400 miles from their source at speeds reaching 55 knots. Telegraph cables on the seabed were broken, one after another, as they lay in the path of the flow.

17. *Measuring the Movements*

Ocean currents play important roles in many of the activities of man. The understanding of both long- and short-term fluctuations in climate and weather requires an understanding of the ocean circulation. Heat from the ocean provides energy for the atmosphere, which, in turn, drives the ocean currents. The current pattern, in turn, determines where and to what extent heat is released to the atmosphere. This feedback system is complicated by the fact that cloud cover over the oceans also determines to what extent the ocean is heated. Ocean water is heated at low latitudes and cooled at high latitudes. But the circular current pattern results in the nearshore water in the middle latitudes becoming warmer on the western margins of the ocean than on the eastern shores, while in higher latitudes the opposite is true. The climate of the Norwegian coast, for instance, is far more pleasant than that of Hudson Bay. The climate effects of currents are not confined to mankind. They are responsible for the distribution of life in the sea, which is dependent upon both temperature and salinity.

Of less direct importance to the average person but of great importance to commercial sea traffic is the effect of currents upon navigation. The effect upon the speed of a ship can be very considerable in the case of the western boundary currents, such as the Gulf Stream, which can make as much as 100 miles difference in the distance traveled by a ship in 24 hours. Subsurface currents, such as the one in the Straits of Gibraltar, have enabled submarines to drift through the Straits undetected, with silent motors.

Icebergs have their origin on land and are carried into the shipping lanes by currents. Most of the icebergs that endanger North Atlantic shipping are calved on the west coast of Greenland, from which they drift with the West Greenland Current and then through the Davis Strait to the Labrador Current. Several hundred every year reach as far south as latitude 48 degrees.

The geologist is concerned with the role of currents in transporting the sediments that cover the sea floor. Other scientists are concerned about the role of currents in dispersing waste material or accidental pollution. The biologist is concerned with the dispersal of waste products and the transport of eggs and larvae and such vital necessities as dissolved oxygen and nutrient materials. The latter is of special concern to the fishing industry.

As living animals and plants in the ocean die and decay, they sink slowly toward the bottom. When finally decomposed, inorganic materials, nitrates, and phosphates are released. If there were no vertical circulation in the ocean, there would be a continual loss of nutrient from the surface and an accumulation on the bottom. However, as explained earlier, at certain places, the deep, cold water rises to the surface and carries with it the accumulated dissolved nutrients. Wherever this process takes place the rich nutrient content supplies the condition for a luxurious growth of phytoplankton, which, in turn, provide for the larger organisms upon which fishes or whales feed. This growth occurs off the coasts of California, Peru, and Chile, and the west coast of Africa. Similar effects take place at divergences, such as in the Antarctic Divergence.

The effects of upwelling are not always beneficial, as in the occasional years when the plankton becomes excessively concentrated or contains noxious species. When this happens, the surface of the ocean may be discolored and fishes killed. Such happenings are often called "red tide," because of the discoloration.

Because of the important role currents play in human life, a knowledge of how to measure them has become vital. This can be carried out directly in the case of surface currents, but not in the case of vertical and deep currents. The simplest and oldest method is by comparing the dead reckoning of a ship's passage, that is its course and speed through the water, with the actual progress made over the bottom.

Measurements by instruments are accomplished in two ways, depending on whether it is desired to measure the rate of flow and direction and their variation with time and depth at a fixed place, or to measure the course directions and speeds of an object moving with the current. The first measurement is known as Eulerian, the second Lagrangian, after mathematicians who developed theories of fluid flow.

The Eulerian measurement may be accomplished by an instrument fitted with a propeller similar to that of an anemometer and a counting device to record the revolutions over a given period of time. Several instruments may be attached to a weighted cable leading from an anchored vessel or other stationary object. Each instrument has a vane to orient it in the path of the current and devices to record the current direction. In this way the currents at different depths are simultaneously recorded. One of the earliest meters of this type, named for its inventor, V. W. Ekman, has an ingenious device for recording current direction. At each revolution of a mechanism geared to the propeller shaft, an opening is uncovered, allowing a small pellet to drop from a container into a small cup mounted on a compass. The pellet runs out of the cup along a groove mounted along the compass needle, which guides it into one of thirty-six radial chambers of a circular tray, each chamber representing a 10-degree angle. As the current changes direction, the pellets drop into one or another chamber, thus recording the amount of current flow in each direction by the total number of pellets in each chamber.

While some Eulerian measurements may be recorded in the instrument, other devices are equipped to transmit the information by electrical cable, radio, or ultrasonic sound. In place of propellers, some instruments are arranged to measure the pressure of water flow exerted upon a fixed plate or the slope of a wire suspended from a fixed point with a drag attached to the end of it. Still others measure the change in the speed of sound propagated from one part of the instrument to a receiver at a fixed distance. Since the speed of sound depends upon the relative rate of flow of the water, the current will be measured in microsecond time intervals.

The Lagrangian measurement may be made in a number of ways. One of the simplest is to release at designated points in the ocean bottles or waterproof plastic envelopes containing numbered cards with return addresses. If and when these are returned, by the persons who eventually pick them

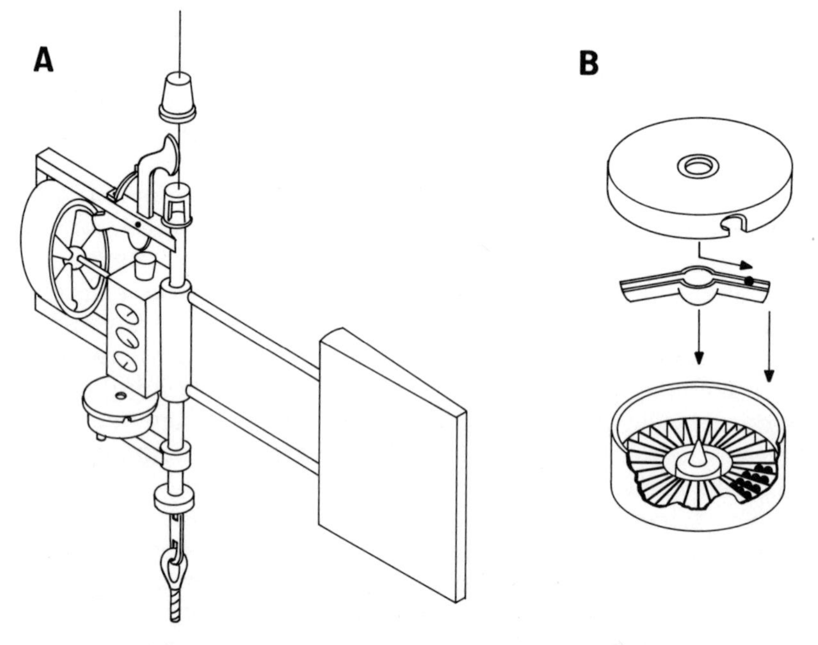

Water currents at points from the sea surface down are measured by means of propeller-driven meters, suspended from an anchored, stationary ship or buoy. In A, *an Ekman meter is shown suspended from a vertical cable. A series of these may be attached to the cable at various depths and their operation triggered by messenger weights sliding down the cable. The meter is held in the direction of water flow by the vertical fin at the right. At the left, the propeller is turned by the water flow and the total flow is indicated by the dials that the propeller drives.* B *illustrates an arrangement in which small pellets are released at intervals by the rotation of the propeller mechanisms into a groove in the tip of a magnetic needle. As the direction of the water current changes, the pellets are distributed to different compartments in the box below, each compartment corresponding to a compass direction.*

up at sea or on a beach, with information as to place and date of recovery, some idea as to maximum time of drift and place of origin and recovery is obtained. Drift bottles made slightly negatively buoyant and with a wire "feeler" may be made to drift along the bottom, with the wire just touching the sea floor.

Chemicals, either dyes or radioactive materials, have also been used. After being released, their subsequent dispersion by currents may be followed by ships or by aerial photographs. Dyes have been used to study the finer structure of currents and eddies beneath the surface, with divers making observations.

Floats equipped with radio have been used. They may be tracked by aircraft, ships, or land-based stations using a radio direction-finding receiver. Some of these floats are made vertically buoyant for a particular depth of water and are tracked by means of a sound emitter attached to the float. The sounds are detected by a hydrophone aboard ship.

One of the problems of surface floats is that the direct action of wind stress may produce a movement of the float in addition to the movement of the water itself. To correct this a float may be spar- or pole-shaped, with most of its surface underwater, or a sea anchor may be submerged with a small surface buoy to mark its position.

There are a number of indirect methods that have been effectively used to study current speeds and volume transport. Some of these depend upon the relationship between density, distribution, and velocity. Others depend upon electrical voltages induced by the passage of salt water acting as an electrical conductor as it passes the earth's magnetic field.

Since the western portion of gyral circulations transports heat from the lower latitudes to the higher latitudes and the eastern portion carries cooler water toward the equator, it follows that the existence and strength of these currents can be inferred when the isotherms, or lines, of equal temperature are distorted, that is, when they point in the direction of the current instead of being parallel to the lines of latitude. In the case of subsurface currents a similar study of the distribution of salinity and temperature in three dimensions gives valuable clues to the nature of these movements.

When a current is flowing steadily, under conditions of equilibrium, the Coriolis force is balanced by the pressure gradients within the water and these, in turn, are dependent upon the density distribution. The warmer, lighter water rises to the right of the current, facing the direction of flow in the Northern Hemisphere so that the surface of the water slopes up to the right. Generally, since the slope of the surface

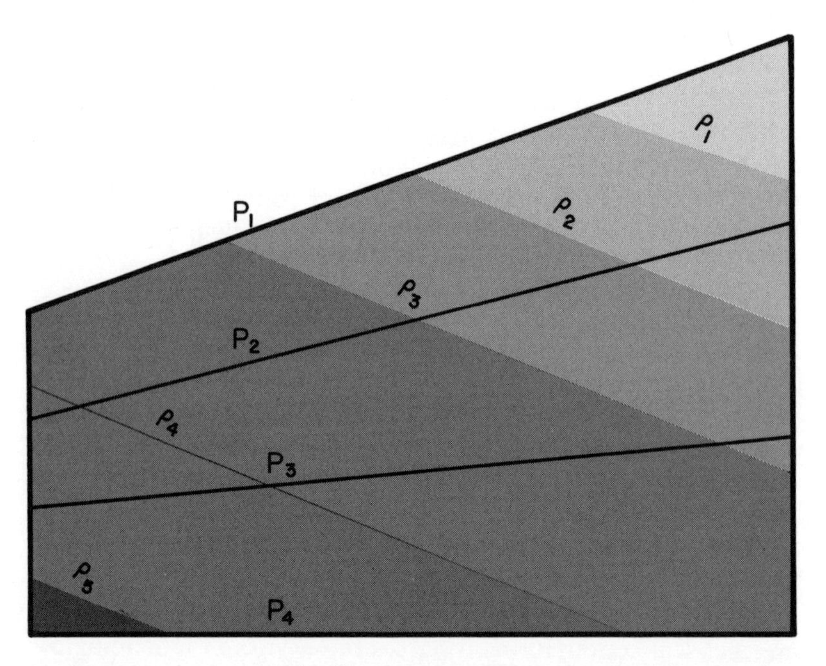

Ocean currents may be measured indirectly by means of the distribution of density within the moving water. This diagram shows a vertical section through the Florida Current, looking northward in the direction of the Current. The surface of the water slopes upward from the Florida boundary on the left to the Bahamas boundary on the right. The slope is proportional to the rate of flow and amounts to 2 or 3 feet over the 40-mile breadth of the current. The slope itself is difficult to measure directly, but it can be calculated from the density distribution. P_1, P_2, P_3, *and* P_4 *indicate the surfaces of equal pressure, while* ρ_1, ρ_2, ρ_3, ρ_4, *and* ρ_5 *indicate the surfaces of equal density, with the warmer, lighter water at the upper right and the heavier, colder water at the lower left.*

is small, it is not easy to make direct measurements. The slope across the Florida Current from Florida to the Bahamas, for instance, is about 2 feet in 42 miles. Nevertheless, ingenious methods of direct measurement have been attempted. A resourceful Danish scientist laid pipes across the Great Belt, between the Danish islands of Zealand and Fyn, a distance of 12 miles, and across the sound between Denmark and Sweden, a distance of 3 miles. Water in the pipes must reach a balance with the same level at each end, except for temperature corrections, and can therefore be used as a reference level for observing the sea level differences.

More often, the cross-current slope is measured indirectly. At some depth below the surface, where no current is flowing, the isobars must be level, since without current flow there will be no adjustment of the isobaric surfaces. It follows that the weight of a column of water above this level will be the same everywhere across the current. The surface slope will therefore reflect the different densities of water beneath the current at different places. This can be measured, using sensitive thermometers to measure temperature at different depths and sample bottles to retrieve water for salinity measurements.

The earth's magnetic field makes it possible to obtain continuous measurements of the current flow while a ship is underway, no matter what the speed of the ship. The principle is the same as that of an electric generator. When an electrical conductor is moved across a magnetic field, electricity is generated. When an ocean current flows across the earth's magnetic field the same effect is produced. Thus, a ship towing two electrodes behind it, at some distance apart on a cable, is able to measure the amount of electricity generated, which, in turn, is a measure of the amount of current at right angles to the ship. By repeating the measurement on a heading differing by 90 degrees, the remaining component of the current is found. Electrical measurements of currents may also be made by means of fixed electrodes on opposite sides of straits.

From the point of view of the yachtsman or the ocean-racing sailor, the characteristics of ocean currents provide some clues as to whether, during the Bermuda Race for instance, the ship is in the current or not. The temperature rise upon entering the current is appreciable. Sometimes there may be definite color changes at the western edge, but this is not always well marked. When there is a rapid change from little or no current to a strong current the most obvious indication may be a change in the surface action of the sea, as waves become steeper in an adverse current and smoother in a favorable current. In the Florida Straits, for the Southern Ocean Circuit Races, there may be more concern about the effects of local winds upon the current and the development of inshore countercurrents. It is also known that there is a tidal periodicity in the flow of the current. While these factors are not too well understood, the information given in earlier chapters should be of help in estimating any departure from average conditions.

For practical information on surface currents, the yachtsman will find the Pilot Charts published by the U.S. Navy Oceanographic Office of inestimable value. These are issued for each month of the year and give the average speed of the currents as well as winds.

18. Power from Sea Motion

Those who have experienced the fury of a fully developed sea in gale force winds or even those whose acquaintance with the energy of waves, tides, and currents is limited to the breakers on the seashore must at some time have pondered over the possibility of harnessing these movements for useful purposes. Surely such gigantic waves as the 110-foot wave reported by U.S.S. *Ramapo* or those that destroyed not once but several times the Minot Ledge and Eddystone lighthouses could be used to generate electricity? So, it seems very reasonable to inquire why we have not begun to replace the "dirty" heat sources of power with ocean power.

Power is the rate at which energy is developed. The amount of potential power in a wave train breaking upon the shore therefore depends upon the energy in a single wave and the rate at which the waves arrive, that is, their frequency. The energy in a wave consists of two parts. One is the potential energy resulting from the height of the wave, the other is the kinetic energy locked up in the orbital movements of the water particles. The potential energy may be considered as the amount of work necessary to take the amount of water from below sea level needed to create the trough and to lift it above sea level to form the crest. The total energy is double this amount, since one must also include the kinetic energy. Once this energy has been determined for a wave train of a specific wave height, it is then necessary to multiply it by the number of waves that arrive per hour to estimate the horsepower or number of kilowatts generated per foot of coastline. To take a concrete example, a 6-foot swell with a period of 16 seconds could generate along every yard of oceanfront about 40 kilowatts. A mile of oceanfront would generate 70,000 kilowatts or 70 megawatts.

There are a number of reasons why there does not exist any large-scale power plant using wave energy. Among the prime reasons is that wave energy is not reliable. It rarely continues to dispense power of the same magnitude day after day in any one place all year long. Another is that, except in exceptional gales, the energy is thinly distributed and to be of practical value it must be concentrated. Still another is that where the power generated is greatest it is not easily controllable, that is, it is not feasible in those areas to erect a machine in which the power could be used to turn a dynamo. In most of these respects, the problems of harnessing wave power are remarkably similar to those of harnessing solar power. Many attempts have been made to harness solar energy, either as a heat source for generating electricity or for distilling freshwater from salt or brackish water. But even where there is a maximum of daily hours of sun, the amount reaching a unit area of land surface is such that a huge collection area must be used in order to provide any substantial amount of power.

Some promising attempts have been made to provide small wave-power motors. Some are based upon the vertical motions of a floating machine, which by suitable mechanical linkage or pumping devices can operate a small generator. The

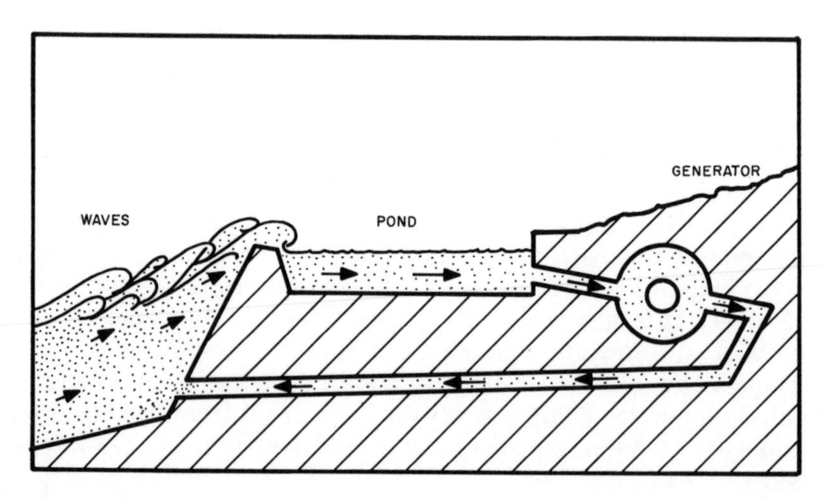

Electrical power might be generated from wave energy by trapping the water of large breaking waves in a raised basin, from which it could then run back to sea level through a suitable hydroelectric plant.

signals of a whistle buoy can be driven by wave motion, or a small amount of electrical power can be obtained. On a larger scale, one of the most interesting proposals is that of a coastal engineer, John Michel, who has suggested that where there is a lagoon, separated from the sea by a strip of land, the breaking waves, in their forward surge, could be led to spill over a sloping ramp so as to collect water in the lagoon at wave height. The entrapped water, at a height above sea level equal to half the wave height, could then be allowed to run back to the sea through a hydroelectric plant. Where a natural situation such as this exists the scheme could well be practical. But where the water breaking ramp needs to be built and maintained, a considerable area of construction would be required, and the low head of water would not be sufficient to provide the substantial amount of power that would justify the scheme economically. Where waves are very high, in the order of tens of feet, the cost of a sufficiently strong ramp and enclosure and its maintenance would probably be unacceptable, while the regular and continuous arrival of such waves would be unlikely.

The long waves of the tides offer a more practical source of power. In fact, tidal power was used in Europe to grind grain as early as the eleventh century, and mills of this kind were still being built in the early nineteenth century in both England and the United States. Some of these merely allowed the rising tide to move the machinery by flowing through a paddle system and the falling tide to do the same in reverse. Only a few stored water at high tide behind a dam and so provided the maximum head of water at low tide. Until recently, a 200-year-old tidal mill was still operating at Slipper Mill, England.

The difficulties in utilizing tidal power are less than those encountered in trying to harness waves. The amount of power generated in a hydroelectric system depends upon the volume of water flowing and the head, or total height through which it drops. In most places the tidal rise and fall is very small compared to that available in the major hydroelectric projects. The volume, too, is inadequate to compensate for the small head. In any case, a small head, even with a large

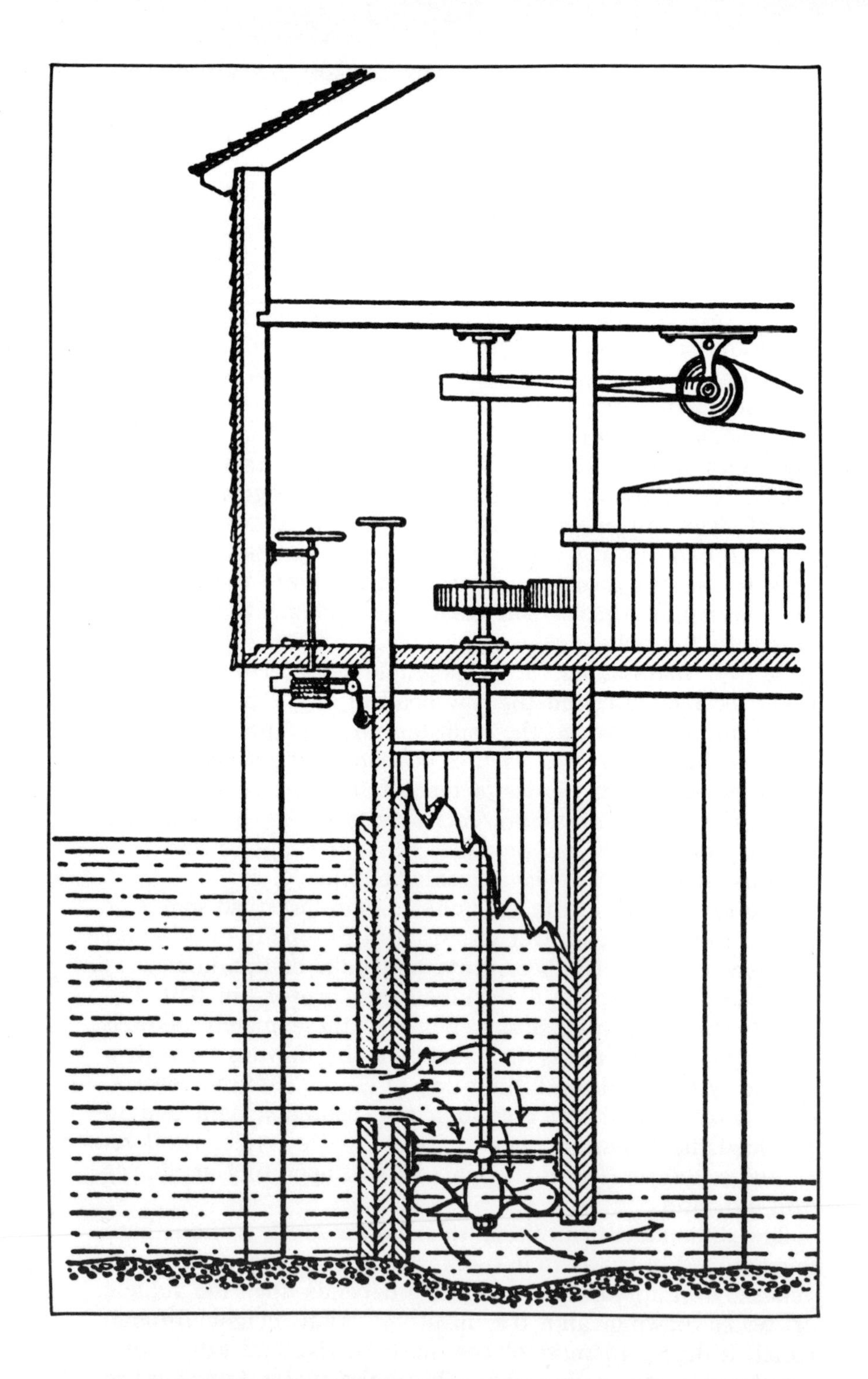

More than 200 years old and still running, Slipper Mill in England is activated both by the tide and by water flowing from a small stream. (Rex Wailes)

flow, requires a far more massive turbine, just as a slow-moving engine requires a larger propeller to drive a boat than a faster engine of similar power.

For at least forty years engineers have looked with interest at the Bay of Fundy as a suitable place for a large tidal plant, since the world's greatest tides occur there. But there has never been complete agreement as to the economic feasibility of operating a plant there. One of the problems in tidal power plants is that the source of power fluctuates. Without dams, the maximum power is generated at mid-water and zero power at slack water high or low. With dams to trap water in a reservoir at high-tide level and to hold water at low level in another reservoir, more effective use can be made. By running water through the power plant from the high reservoir to the low one as the tide rises, and then to the open ocean as the tide falls again, the fluctuations in power are reduced, but there still remains, under the best conditions, a ratio of two to one between maximum and minimum power daily. The periods of maximum and minimum output also change daily, as the daily tidal cycle moves back 50 minutes each day. There are also the fluctuations between spring and neap tides and the longer cycles described in earlier chapters. The cycles of tidal maxima do not coincide with the daily and seasonal cycles of demand for electricity. Only a costly system of storing energy in some form, utilizing auxiliary pumping systems and reservoirs much like a storage battery that is charged when power demand is low and discharged when demand is high, can gear the power supply to the fluctuating power demand.

In the late 1930's the perennial interest in tidal power from the Bay of Fundy was revived and three dams were actually built, connecting the mainland to Eastport, Maine, and separating Cobscook Bay from Passamaquoddy Bay, before the project was abandoned. Once more interest was aroused in 1956 and in 1959 the United States and Canadian International Joint Commission received technical reports recommending a plan. Under this plan, Passamaquoddy Bay was to be made a high water-level pool and Cobscook Bay would become a low-level pool, through appropriate dams. Water

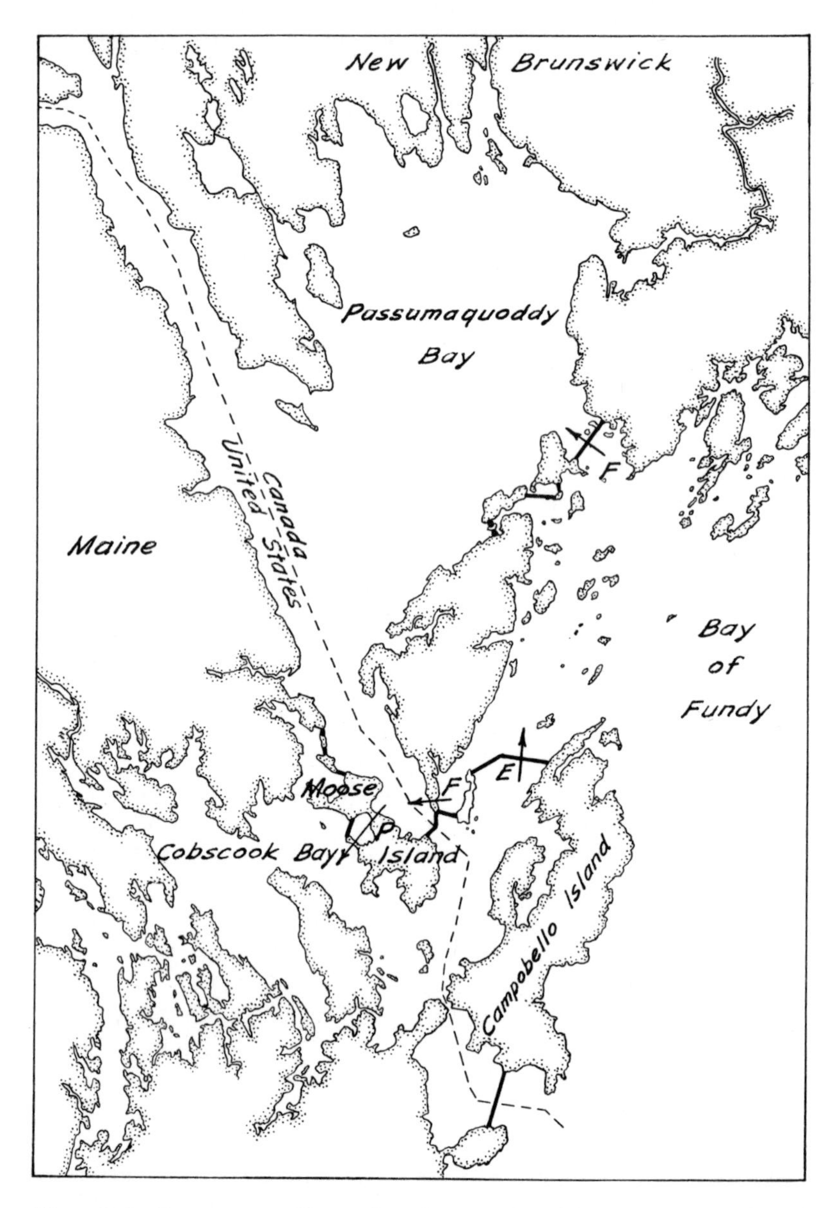

Map of the Passamaquoddy project. Passamaquoddy and Cobscook bays would be dammed to form high and low pools. Filling gates are designated by "F," emptying gates by "E," and the power unit by "P."

flowing from the high-level pool to the low-level pool would drive the turbines. A number of sources of auxiliary power were proposed. One would pump water from Passamaquoddy Bay into the Digdeguash River estuary and around to form a reservoir at the northern end of the bay. Another would provide a second hydroelectric generator at a dam across the St. John River at Rankin Rapids.

Built behind a wall of cofferdams, the Rance power plant required the drying of 190 acres of estuary bed. The view is from the left bank, facing the town of St.-Servan-sur-Mer. (Brigaud, Electricité de France)

The Passamaquoddy project involved a careful study of the possible environmental effects, especially upon the fisheries of the area. Mean water level would be raised about 6 feet in Passamaquoddy Bay and lowered 5 feet in Cobscook Bay. Because of the dam gates, closed except for about one quarter of each tidal cycle, the currents would be altered and the surface waters would become less saline and warmer in the summer but more likely to ice over in the winter. Although the clam fishery would be hurt, the changed conditions would be beneficial for flounders, scallops, lobsters, salmon, alewives, and striped bass. However, the whole exercise was in vain, for the project was again abandoned. It would have cost about $600,000,000, shared between the United States and Canada. In 1964, however, a new study again recommended the two-basin plan with a total tidal power production of 1 million megawatts—but still no progress.

More encouraging than the Passamaquoddy scheme is a relatively modest plant that is actually in operation and producing about 500,000 kilowatts in Brittany. The Rance River plant near Saint-Malo is situated at a place where the tidal range is between 30 and 47 feet. The dam across the river is about a half mile long with a 13-mile natural basin. A feature of this plant is the reversing turbines. These generate electricity when driven by the flow of water, but pump water when fed electricity, thus enabling the net electrical output of the system to be evened out or tuned to demand. Power is generated on the incoming tide and also when the flow is reversed, at a rate of 24,000 cubic yards per second. But at high tide, power from the national electric grid is used to pump still more water upstream, so that an extra 3 feet of head is gained. At low tide water is pumped from upstream, thus further increasing the effective range. The cost of harnessing this tidal power was $100,000,000.

There are other places where tidal power plants are feasible or actively planned. At least ten places in Great Britain have tidal ranges in excess of 30 feet. Among these are two of the British Channel Islands, Jersey, and Guernsey. In France more than twenty localities have similarly large tidal ranges; Brazil has one and Argentina has three. Two good locations exist in Australia. Larger tidal ranges are found in Korea, and in India 40-foot ranges occur in certain areas. Several places in Canada have equally large ranges.

There are plans for a tidal power plant of about 800 megawatts in the Severn River in Great Britain, where the 46-foot tides would justify it. Because the tides at the Severn and the Rance differ in phase, that is, one is high when the other is low, an electrical linkage of the two would smooth out production peaks. Russia has already built an experimental plant and has planned a considerable number of tidal dams for the White Sea.

Tidal energy production, though small compared to that of hydroelectric, nuclear, and fossil-fuel plants, is nevertheless a tribute to man's slow understanding of and ability to predict and control the tides. But the harnessing of ocean currents, some of which amount to thousands of times the flow of tidal streams, is far away. Although there has been considerable speculation regarding the utilization of major ocean currents and the probable effects upon climate, no plans have been drawn up, and speculations, for the time being, are far from being realistic.

Bibliography

Representative, but not comprehensive. Selected for availability and range of level from popular to advanced.

Barber, N. F., *Water Waves.* London and Winchester, England: Wykeham Publications, 1969. Paperback. Interesting mathematical treatment. Elementary level.

Bascom, W., *Waves and Beaches.* Garden City, N.Y.: Doubleday Anchor Books, 1964. Good elementary paperback.

Bird, E. C. F., *Coasts.* Cambridge, Mass.: M.I.T. Press, 1969. The effects of wind, wave, and current upon coastlines. Elementary.

Coles, K. A., *Heavy Weather Sailing.* Tuckahoe, N.Y.: John de Graff, Inc., 1968. Outstanding account of heavy weather, phenomenal waves, and their effects upon ocean racing yachts. Analysis of wind and wave conditions leading to loss or damage to sailing vessels and precautions that might have minimized their effects. Highly recommended to yachtsmen and sailors.

Duxbury, A. C., *The Earth and Its Ocean.* Reading, Mass.: Addison-Wesley Publishing Co., Inc., 1971. Waves, tides, and currents. General student level.

Groen, P., *The Waters of the Sea.* London: Van Nostrand, 1967. Waves, tides, currents, and sea ice. General reader or elementary student level.

Hidy, G. M., *The Waves.* New York: Van Nostrand Reinhold Co., 1971. Mostly about currents and elementary fluid dynamics. Mathematical.

Ippen, A. T., ed., *Estuary and Coastline Hydrodynamics.* New York: McGraw-Hill Book Co., 1966. Action of waves and currents upon coastlines and estuaries. Advanced level. Mathematical.

Johnson, D. W., *Shore Processes and Shoreline Development.* New York: Hafner Publishing Co., Inc., 1965. Standard text.

Kinsman, B., *Wind Waves.* Englewood Cliffs, N.J.: Prentice-Hall, Inc., 1965. Advanced student level. Mathematical.

Moore, H. B., *Marine Ecology.* New York: John Wiley & Sons, Inc., 1958. Contains information on life in the intertidal zone. Student level.

Neumann, G., *Ocean Currents.* New York: American Elsevier Publishing Co., Inc., 1968. For the physical oceanographer. Graduate level. Good bibliography.

Pickard, G. L., *Descriptive Physical Oceanography.* Elmsford, N.Y.: Pergamon Press, Inc., 1964. General. Coastal processes, currents, and water budgets. Undergraduate level.

Russell, R. C. H., and D. H. MacMillan, *Waves and Tides.* London: Hutchinson Scientific and Technical Publications, 1954. Elementary level or general reader.

Steers, J. A., *Coasts and Beaches*. Edinburgh, Scotland: Oliver & Boyd Ltd., 1969. Good paperback for elementary student or general reader.

Stommel, Henry, *The Gulf Stream*, 2nd ed. Berkeley, Calif.: University of California Press, 1965. The Gulf Stream System. Intermediate student level.

Sverdrup, H. U., M. W. Johnson, and R. H. Fleming, *The Oceans*. Englewood Cliffs, N.J.: Prentice-Hall, Inc., 1942. Good standard text of general oceanography.

Tricker, R. A. R., *Bores, Breakers, Waves and Wakes*. New York: American Elsevier Publishing Co., Inc., 1965. An excellent book. Elementary mathematics but of interest to general reader.

Wiegel, R. L., *Oceanographical Engineering*. Englewood Cliffs, N.J.: Prentice-Hall, Inc., 1964. Coastal engineering. Advanced level.

Index